PRECURSORS

The Human Ancestry

Landon S. Hayes

Dedication

This book is dedicated to my family and my best dog, Buddy, who recently passed away. Without them, none of this would've been possible.

Landon S. Hayes

Introduction

Have you ever wondered about the incredible stories you discovered that you can go back to with your ancestors? Not like here, not just a few generations ago, but through centuries, to a large number of ages prior, to an ancestor who shared the world with the Neanderthals. Return 60,000 ages and we will meet one of the first people to walk outside Africa. Meeting 100,000 generations and relatives would be scarier than you think.

    Well, that's what we're going to do— meet the creatures that have changed us and see how their lives have led to our wonderful reality. Our long-dead progenitors left a key to open the past. With their bones and relics as a guide, scientists can create an image of what it would be like to go back in time. These little pieces hold secrets not only about how our ancestors died but also about how they lived. From our closest relatives to Neanderthals to the first fearsome Europeans. From the ancestors with the ingenuity to escape from the rainforests of Asia to the first real families who worked together in the arid environments of Africa.

We travel through time, revealing the mysterious ingredient that makes us unique at every step of the way, meeting the first creatures that have them— the early toolmakers, the heavyweight rivals who shaped our past. We go back millions of years to meet a creature that led us on a journey from apes to us.

About Human Evolution

We humans like to think of ourselves as distinct, creatures that are above the rest of the animal kingdom. But we have understood our position in Natural History since the 19th century. We are just another mammal, connected to a beastly history by our family tree. Yet one of the most fascinating of evolution is the story of how we became so special.

It is easy to understand why we feel so isolated. Humans are different from all animals. However, humans are animals, a species of big mammals. In this case, there is a difference between us and all other species which leads us to call them animals and view them as separate from us. We think that centipedes, chimpanzees, and clams share some animal characteristics that we don't have, or that we have human characteristics that they don't have. Such human features include language-based speech, making art, making complex tools, wearing clothes, and darker features such as attacking our own and other species in mass numbers. Though, we inherit the same parts of the body such as molecules and genes as do other animals. What kind of animal we are is clear.

We belong to a family called the primates, which are typically tree-dwelling animals with forward-facing eyes and keen eyesight, long arms, and gripping fingers, and opposable thumbs. Most of them have very large brains and a high level of intelligence and live in social groups. There are two suborders: anthropoids, the advanced primates, which include monkeys, apes, and humans; and prosimians, the primitive primates, which include lemurs, tarsiers, aye-ayes, and lorises. The first primates were minuscule, squirrel-like animals that scurried in the trees at around the same time the dinosaurs died out.

Unlike Hollywood movies that depict early man living along with dinosaurs, our ancestors did not live in the same time as the dinosaurs or even hunt them. The last dinosaur died off 66 million years before man. Therefore, humans were off the dinosaur's menu.

After the dinosaurs died out, primates started out as being nocturnal animals, living in the trees in the dark and eating insects on the wing. They were solitary and could only see in black and white. With the dinosaurs gone, primates were able to evolve into many new species, becoming larger and more intelligent over time. We, humans, belong to the same family as apes and are very similar to chimpanzees. We share about 99% of our DNA with chimpanzees, making them our closest living relatives. While both lemurs and monkeys have tails apes don't. But they wouldn't require tails because they were built for life on the ground. But some apes can still climb in the trees with ease like the gibbons or chimpanzees.

  Every extinct and living species on earth have resulted from the same evolutionary processes determining the way they are through shaping their morphology, physiology, and behavior. The traits specific for the human species are the result of the same evolutionary processes responsible for any other living creature. From a general evolutionary perspective, humans are consequently no different than any other species on the planet.

  For many people, paleoanthropology is an

exciting scientific field because it investigates the origin, over millions of years, of the universal and defining traits of our species. Yet the exact nature of our evolutionary relationships has been the subject of debate and investigation since the great British naturalist Charles Darwin (1809-1882) published his monumental books On the Origin of Species (1859) and The Descent of Man (1871). Darwin never claimed, as some of his Victorian contemporaries insisted he had, that "man was descended from the apes," and modern scientists would view such a statement as a useless resolution—just as they would dismiss any popular notions that a certain extinct species is the "missing link" between humans and the apes. There is theoretically, however, a common ancestor that existed millions of years ago. This ancestral species does not constitute a "missing link" along a lineage but rather a node for divergence into separate lineages. This ancient primate has not been identified and may never be known with certainty, because fossil relationships are unclear even within the human lineage, which is more recent. In fact, the human "family tree" may be better described as a "family bush," within which it is impossible to connect a full chronological series of species, leading to Homo sapiens that experts can agree upon. However, some people find the concept of human evolution a bit troubling because it does not seem to fit with religious and other traditional beliefs about how people, other living things, and the world came to be. Nevertheless, many people have come to reconcile their beliefs with the scientific evidence.

Evolution is sort of our history. Without it, we wouldn't know our origin of where we came from or how or why we became this way. As the centuries wore on, we became the dominant species on the planet. We built great cities and built weapons from stone tools into the most powerful advanced ones than ever before. We fought wars against each other and we even been through hard times. We also invented religious and cultural customs.

    We humans have really come a long way. If our ancestors couldn't make tools or make fire, they wouldn't survive an ice age or defend against predators. But thanks to these abilities, we can think and solve problems today. We are now the rulers of the Earth.

"It is the nature of man to rise to greatness is expected of him."

—John Steinbeck

"Struggle is the father of all things. It is not by the principles of humanity that man lives or is able to preserve himself above the animal world, but solely by means of the brutal struggle."

—Charles Darwin

Before the Dawn

Period: Early Neogene (Late Miocene, Early Messinian)

Date: 7,000,000 BC

Location: Eastern Africa

The green-filtered light of the forest cast a ghostly glow on the thick, twisting ropes of lianas that snaked between the trees. Birds darted swiftly among the gaps in the canopy, chasing insects on the wing, silent except for the noise of their wings as they whipped through the air. Giant spider webs, white and gauzy, hung between the branches like ruined sails. The air was damp and humid, the thick beds of leaf litter giving off a strong, earthly smell. The forest was alive with rustlings in the treetops and strange howling-hoot calls, but it was impossible to see the creatures that were making them. Clearly, this was not a world of Ardipithecus ramidus, but we have traveled only in time, not space. This was Africa, but 4 million years before the genus Ardipithecus existed. Then an unbroken sea of primary rainforest stretched across Africa, extending for hundreds of miles in all directions. This type of forest-covered many other parts of the world too, since the planet was warmer than it is today and capable of supporting this more luxuriant vegetation.

This was a world in three dimensions. The animals that lived here were as happy 300 feet up in the air as they were on the ground, or happier, perhaps since many unseen dangers lurked on the forest floor. These were animals that could move as easily up and down as they could from left to right. Many of the simian primates that inhabited this forest were very much like the simians of the future Quaternary period, but there was also a great ape species here that was unlike many others.

In this ancient forest lived the somewhat rare ape known as Sahelanthropus, the distant ancestor of Ardipithecus. Little was known about how it looked and lived, but it indeed existed in this immense forest. It was possibly very close to the time of the chimpanzee/human divergence.

This mammal was not entirely bipedal, mostly a knuckle-walker, supporting its upper body on the backs of its hands. Resting on the knuckles held the body semi-erect when the animal was walking on the ground, whereas for a truly quadrupedal animal, the body was horizontal. Since evolution works by tinkering with the body structures that already exist, rather than evolving new ones completely from scratch, an ancestor that was already half-upright would be the simplest path for evolution to take.

Alternatively, Sahelanthropus was also a small more brachiate, using its arms to swing through the trees, like gibbons. Gibbons always walk on two feet on the rare occasions that they come to the ground, their long arms held above their head in a fashion that seems rather comical, but such a strategy may partly help

Sahelanthropus to make the switch from the trees to the ground.

One question that still remains unanswered is why these changes happened in the first place. What caused the shift to a bipedal life on the ground? The answer lies not in the African jungles, but half a world away, with a series of events that altered this planet forever.

Although the ground appeared stable and solid, appearances were deceptive. The planet's surface was dynamic and constantly changing. Beneath the surface, Earth had a boiling, churning center of molten rock, known as magma. The thin crust, on which all life forms lived, was also less stable than one can imagine. It was not continuous but formed a dozen or so large plates, which rode on top of the magma and moved against each other constantly. Throughout Earth's vast history, these continental plates had joined up and separated, buckling and contorting the huge landmasses that lay on top of them, altering the planet's climate and having dramatic effects on its plants, animals, and other life forms.

In a process beginning in the Late Cretaceous period, during 75 million BC, these movements of the continental plates gradually shunted India (which at the time was a massive floating island subcontinent) into the main body of the Asian continent. This clash of landmasses pushed up a vast formation of rock, forming the largest range of mountains on Earth: the Himalayas.

At 29,029 feet in elevation and varying in width from 250 miles in the west to 93 miles in the east, the emergence of this enormous range had an extraordinary effect on the planet's atmosphere. Moist air drawn from the Indian Ocean was no longer able to get past the vast barrier formed by Himalaya. Instead, the clouds were forced to drop their load over India, causing some of the heaviest rainstorms the world has ever seen, later on, called the monsoons. These immense rainstorms meant that the air that continued westwards lacked moisture. As it passed over the Middle East and on to East Africa, it started to have a drying effect. Although the story is complicated, many think that, in the Early Neogene period (specifically during the Late Miocene epoch) around 7 or 6 million BC, a threshold was reached when the climate changed sufficiently to cause a shift in the types of plants that could survive in these areas. The moist forests that once flourished in East Africa gradually began to dry up and die out.

But the formation of the Himalayan Mountains and the monsoons was not the only event that helped to reshape the African continent. Around 25-22 million

BC, in the Late Paleogene to the Early Neogene (during the Late Oligocene to Early Miocene specifically), the margin of the continental plate that lay below East Africa began 'doming', with huge uplifts of lava being pushed up from the gap between the plates and creating uplifts over 30,000 feet in height. Weakened by the separating plates, the continental rock gradually began collapsing in a long, vertical fault, several thousand kilometers long and up to nearly two miles deep. It is now known as the East African Rift. The effect of this huge tear on the world's surface was to throw the eastern side of the continent into rain shadow. This deprived it of moisture and further exaggerated the drying effect induced by the monsoons in Asia.

To the west of the continental rift zone, there was little or no change; the rainforests continued to receive enough moisture to survive, and the animals that lived in them continued on the evolutionary path that led to such later species as chimpanzees and gorillas. But on the eastern side of the rift, it was a different story as huge swathes of the forest died back, leaving a mosaic of swamps, brushy woodlands, and open grasslands. Forests could only occur along the margins of rivers where there was still sufficient water to support large trees.

For the creatures caught in this great transformation, the pressures to adapt were enormous; the specializations that made each species successful in a dense forest environment no longer worked in a more open, patchy habitat. Vast changes in anatomy and behavior were needed to exploit this new world effectively. In all, it took 3 million years to change a forest-living great ape into one that could cope with life in the woodlands of the new East Africa. It was a new kind of great ape altogether (an upright ape, with strong legs designed for walking on the ground), a species called Ardipithecus.

So why exactly did these particular apes start walking upright? Why did this single adaptation make them so successful? One theory is that the advantage of standing upright is related to the way that it allowed Ardipithecus and its relatives to develop new behaviors.

For example, with its hands no longer needed for walking, it was able to carry both food and dependent young more efficiently. Similarly, standing on two legs increased the ape's height, allowing it to peer over tall grasses to spot potential dangers. Walking on two legs benefitted Ardipithecus directly in terms of allowing it to cope with the environmental demands of its new habitat.

This upright success of Ardipithecus (besides their great social bonds and parental care) stems from its unique blend of non-human ape and human characteristics. But it would be a serious mistake to see Ardipithecus as some sort of compromise between the two. It was not an inferior kind of human, nor a superior type of non-human great ape, but a highly successful species in its own right, with a unique and effective way of getting around, which partly helped it from overheating. It was already on the path to true sentience.

First Blood

Period: Late Neogene (Early Pliocene, Early Zanclean)

Date: 4,400,000 BC

Location: Eastern Africa

This area of eastern Africa was a haven for wildlife. It was geographically very active, but the climate supported an abundance of animals, plants, and other organisms. Between the volcanic peaks was a mosaic of grassland savanna, open woodland, and dense tropical forest. This variety of habitats was reflected in the range of animals that sheltered here. Plains creatures such as bovids (like gazelles and impala), proboscideans (like Deinotherium, Anancus, and African elephants), and equids (like Equus and Eurygnathohippus) lived close to forest-dwellers like primates (such as Ardipithecus and Theropithecus), bats (such as Kerivoula, straw-colored fruit bats and mouse-eared bats) and suids (such as Kolpochoerus and Metridiochoerus).

    The African morning was fresh and cool. The sun had not yet risen, but the sky grew pink on the horizon. The insistent calling of a lone diurnal bird waking up was clear and resonant in the still air. As the sky continued to lighten, the dawn chorus began in earnest, crowding the air with birdsong. In the forest bordering the river, a troop of blue monkeys descended from the trees to begin the day's foraging.

But these guenons were not the only primates to have spent the night here. As the sun finally spilled light over the horizon, a much larger simian roused itself and over the next branch from the messy nest of vegetation in which it slept, its relatively long arms supporting its full body weight.

This individual male represented one of the first steps on the journey towards future humanity; he belonged to a species known as Ardipithecus ramidus.

Even though his kinship to future humans was indeed accurate, he looked distinctly like a non-human great ape. His kind's brain was relatively small (~350–300 cm3), like that of a chimpanzee, and possessed the sort of legs and arms that characterize great apes. For example, he had opposable thumbs and toes designed to give him good locomotion in the trees. Those features are not found in later human species. The face was less prognathic but primitive, with a somewhat low forehead and not much of a chin. They ate mainly fruit, vegetables, seeds, arthropods, and, at times, the odd piece of fleshy meat, with the exclusion of nuts and tubers, and while these apes may use tools to extricate termites and ants from their nests, they did not fashion tools from stone or use them as weapons. They communicated with others of their kind with gestures and calls, but they could not speak as their future descendants would. They possessed none of the intellectual and technological advantages that would allow their descendants to dominate the planet.

  High in the trees, the male Ardipithecus was eating leaves from an overhanging his field of vision as he saw a herd of gentle Ancylotherium feeding in the distance. Fully-grown, these animals were 6.6 feet high at the shoulder and weigh 990 pounds, far too big to be threatened by Ardipithecus. Their kind was one of the last of the Chalicotheres, some of whom were giant knuckle-walkers. Ancylotherium browsed on low branches, a bit like caprine bovids, but were otherwise more conventional quadrupedal herbivores.

The Ardipithecus male suddenly heard a noise off to his left. He quickly lifted his head up, his teeth drawn over his lips in a grimace of fear. Using his toes, he carefully walked on the other branch. to see the commotion. Out in the open, two bipedal apes of his kind were running away from a big cat. They ran toward the tree for safety, but one of them didn't make it in time. The ape ended up being the cat's dinner as the saber-tooth dragged the ape away.

Their unfortunate end illustrated, with stark clarity, that Ardipithecus had no special abilities to avoid unexpected predation and other dangers. Unlike the future humans, they did not dominate and control their environment, shaping it to suit them. In fact, they were often at a distinct disadvantage since they were small creatures, usually standing between 3 feet 11 inches to 4 feet 7 inches tall, and they possessed no physical weaponry, such as horns or large and sharp canine teeth, that could be used in self-defense; nor were they fast enough to outrun any danger that arose. For an Ardipithecus of any species, survival was a matter of pure chance. They could not control nature, nature controlled them.

Then, the decision was made, the male in the trees climbed down and dropped gently to the ground and crouched, sniffing the air and scratching through the coarse, dark hair that covered his body. He walked across a clearing, looking out for predation. Ardipithecus were not hunters but gatherers.

Ardipithecus spent most of the day split into small gathering parties. They gathered food for their group to keep them sustained. Small parties were much more efficient all around, although predators, like large felids (Acinonyx, Panthera, Dinofelis, Homotherium, and Megantereon), were more likely to attack small groups. However, their climbing abilities helped keep them out of danger, and the females in particular never strayed too far in the open. Right now, the fig tree they lived in was not bearing any fruit at all. So the gatherers sought out more food. Surely without any supplement of food, they would starve.

A ripe fig tree had drawn this male to the edge of the forest in the northern part of his territory. He gathered all the fruit he could carry. The male paused for a moment, listening to any signs of danger. Clearly, he could not stay for long. The male needed to get out of here before it was too late. He ran as fast as his legs could carry him. Being able to carry things while on the run was another advantage of walking upright. He darted across the clearing and finally made it back to his group. They now had all fruit they needed.

In the trees, the Ardipithecus troop began to climb down. Some decided to stay up in the trees for the time being and while some on the ground sought out grooming partners, presenting a shoulder or an arm in the hope that someone will take up their offer. Not only did grooming remove ticks and lice, but the action of combing through another individual's hair calmed and soothed both the one that groomed and the one that was groomed.

At a base of a fruitless fig tree, one adult female was cradling her infant daughter supported on her chest. She shall be called Ardi. Her casual manner and the ease with which she coped with her wriggling infant show that she was well-versed in the art of childcare and indeed, at 22 years old, she had done this twice before. Her previous offspring were both healthy and still lived in the same troop as their mother, although her elder daughter, at almost 10 years old, would soon reach maturity and leave to find another Ardipithecus ramidus troop in which to live (preventing inbreeding). The bond between these two will continue to develop as the baby grew, and she would always be able to look to her mate for protection against aggression from either predator, something she would need sooner than either of them realized.

Losing interest in food, the young Ardipithecus

pair began some rough-and-tumble games. Enjoying herself immensely, the infant pulled her lips over her teeth and panted at her brother, making a "play face." Ardi's mate came along. Males were the same size as the females since there was less sexual dimorphism in the region. Also, they were monogamous and did not fight for females. Males and females raised their offspring together.

Ardipithecus ramidus society marked the beginning of a kind of relationship that had become crucial to human existence: the pairing up of males and females in partnerships. This community was a smaller tightly-knit complex group, but the consequence of such a network of relationships had been to kick start another peculiarly future human attribute.

To bond with the other members of the community, the Ardipithecus relied on physically grooming each other in the way that the other primates do, there were simply too many individuals for them to have the time to do that. Ardipithecus formed a friendly and intimate relationship and maintained them across a much wider social community. Thus their upright physique, their larger brains, and their more complex social groups had come together to produce this most significant evolutionary step.

The Next Step

Period: Late Neogene (Late Pliocene, Middle Piacenzian)

Date: 3,200,000 BC

Location: Eastern Africa

The scorching heat of midday had died away but it was still hot and very humid. The sun cast down on the rocky African outcrop. At the edge of the cape thatching reeds, where the receding waters had left a bank of dried out mud, a 4-foot-and-6-inch tall male great ape crouched, digging into the mud with a stick. Reaching down into the hole he had made, he pulled up the muddy root of a carex and, after briefly rinsing it in the water, popped it into his mouth and began chewing. Large, strong chewing muscles bulge in his jaw and cheeks as he ground away at his food, steadily doggedly, the mouthful of food taking almost half a minute to consume. His odd, wide face appeared impassive, but he was keeping a close eye on the rest of his troop, who were strung out along the shoreline. Turning his head, he also monitored the few troop members that were digging for food in the shade of the trees that fringed the lake. They also used sticks to dig but, instead of carex roots, blubs and tubers were their reward. Too far from the water to make washing an option, they rubbed their food against their hairy arms to dislodge dirt and grit before eating it. The only sounds to be heard right now were the contented grunts they made at each other, and the sound of relentless chewing, as they pulverized food between their relatively large back molar teeth.

These apes represented one of the first steps on the journey towards future humanity; he belonged to a species known as Australopithecus afarensis. Even though his kinship to future humans was indeed accurate, he looked distinctly like a non-human great ape. His kind's brain was relatively small (~380–430 cm3), like that of a chimpanzee, and possessed the sort of legs and arms that characterized great apes. The face was prognathic and primitive, with a somewhat low forehead and not much of a chin. They ate fruit, vegetables, tubers, seeds, nuts, arthropods, and, at times, the odd piece of fleshy meat, and while these apes may use tools to extricate termites and ants from their nests or crack nuts, they did fashion tools from stone but not use them as weapons (except for maybe a stick used to whack each other during fights if they got the chance). They communicated with others of their kind with gestures and calls, but they could not speak as their future descendants would. They possessed none of the intellectual and technological advantages that would allow their descendants to dominate the planet. But they had the ability to walk upright, and that was enough to allow them to succeed where other apes failed.

In the distance, on the open savanna, a herd of magnificent Deinotherium was destroying a lone acacia tree. These large 11.9-foot-tall proboscideans used their curious downward/backward-curving tusks attached to their lower jaws to strip the bark off the tree and, as if that was not enough, after they had eaten all the bark they could reach they could use their weight and muscular trunks to push the whole tree over and strip the upper branches. Without the activities of Deinotherium herds, the woodlands would be far more extensive.

But the Deinotherium herd was no threat for these great apes unless they went into munst, an act of sexual arousal when male Deinotheres were pumped with testosterone, making them aggressive toward other animals.

Nearby was the burrow of a common dwarf mongoose group and while the neighboring primates fed a hunting party returned. The mongooses rushed around the Australopithecines, greeting and rolling over each other, but ignoring the apes completely. These active little hunters knew exactly what and what not to fear. Gradually they retreated down their burrow, but 5 or 6 remained above on lookout duty.

At that point a young male arrived, cradling an ostrich egg in his arms. Gently he hooted and rolled the egg on the ground, then scampered back and forth to attract attention. The dominant male ignored him, but the others were curious. In particular, one of the females wandered over to sniff the egg. After showing off for a while, a young male came back to the egg, carrying a stone. He thumped at the shell with the stone several times and, having created a small hole, dipped his finger inside to pull out the yolk. The other females abandoned their and joined the young for a taste of egg yolk.

The dominant male was now getting angry. He picked up a stick and dragged it around where Bruiser and the others were feeding. The dwarf mongooses shrieked and disappeared down their burrow, but this was not because of the fight between the two male

Australopithecus. The warning was there, but the apes were too preoccupied to notice. A ten-year-old male was sitting alone by the cross berry as a pale-colored creature, measuring 70 centimeters tall to the shoulder, snuck across the ground towards him. He did not stand a chance.

This felid with canine teeth that were longer and more flattened than those of most other felids was a male Dinofelis, an ambush predator with powerful front limbs and a crushing bite. The ten-year-old male had no more than a moment to cry out before he was bowled over and pinned down. If attacking something as large as an Ancylotherium the Dinofelis might have gone for the throat, but with the ape, he just crushed her skull. It was all over even before the rest of the troop had time to react. Then, with an almost simultaneous screech from four throats, they scrambled for the nearest trees.

Shrieking in alarm and terror they raced about the branches, shaking leaves and staring at the sorry sight below. The Dinofelis adjusted his grip on the child so that he had his neck in his mouth. Then he half lifted him and started to drag him away from his distraught companions.

Big felids were some of the main predators of Australopithecines, especially smaller species, which could climb into trees in pursuit of their prey. Dinofelis usually attacked larger or similar-sized prey (such as Papionins and certain antelope), but if this particular one developed a taste for great apes then the troop was in trouble. The Machairodontine dragged the child's body to a large afzelia tree that leaned over a deep limestone sinkhole and then, using his incredibly strong forelegs and neck, managed to carry him up into the lower branches. Despite his size and power, the

Dinofelis knew that the smell of the kill would attract competition, particularly from hyenas (Pachycrocuta and Spotted hyenas). If he was outnumbered he could be chased off a kill before he had managed to eat his fill. In the tree, he could eat the carcass in peace, or left it and returned to it later until the whole kill was finished or had fallen into the sinkhole.

The child's unfortunate end illustrated, with a stark reminder, that Australopithecus had no special abilities to avoid unexpected predation and other dangers. Unlike the future humans, they did not dominate and control their environment, shaping it to suit them. In fact, they were often at a distinct disadvantage since they were small creatures, usually standing between 3 feet 11 inches to 4 feet 7 inches tall, and they possessed no physical weaponry, such as horns or large and sharp canine teeth, that could be used in self-defense; nor were they fast enough to outrun any danger that arose. For an Australopithecus of any species, survival was a matter of pure chance. They could not control nature, it controlled them.

It had been several days since the Dinofelis attack and the troop was sitting in some lower branches indulging in mutual grooming. Not only did grooming remove ticks and lice, but the action of combing through another individual's hair calmed and soothed both the one that grooms and the one that is groomed. One adult female refuses all such overtures, however, and beings to climb down to the ground, her infant daughter supported on her chest. It is the female who first appeared at the river and witnessed the Dinofelis attack firsthand. She shall be called Lucy. Her casual manner and the ease with which she coped with her wriggling infant showed that she was well-versed in the art of childcare and indeed, at 20 years old, she had done this twice before. Her previous offspring were both healthy and still lived in the same troop as their mother, although her elder daughter, at almost 10 years old, would soon reach maturity and leave to find another Australopithecus afarensis troop in which to live (preventing inbreeding).

The infant female, although much quieter, was still upset, whimpering in distress, her tiny hands gripping her mother's hairy body. Less than a year old, she was growing rapidly and was constantly hungry. Lucy could barely keep up with her daughter's increasing demand for milk and spent virtually all her time feeding to keep her own energy leveled up. She certainly could not afford to hang around too long in the trees this morning; despite the potential dangers down below, she must start feeding, preferably on the best quality food she could find. This did not present Lucy with too much of a problem since, as the most dominant female in the troop, she could easily displace any of the other females. The males, however, were another matter: they weighed almost twice as much as she did and they were considerably taller. She did not stand a chance of competing with them, so she must get a head start and go off alone.

This Australopithecus troop was never very cohesive in any case. The 20 individuals that made up Lucy's troop spent most of the day split into small foraging parties, coming together again only in the evening at the trees in which they sleep. This was the most effective way to exploit their woodland home, with its widely dispersed trees and bushes; it not only reduced competition for food between individuals but also helped to cut down on traveling time. If everyone stuck together, they would need to visit many more food trees each day to make sure they all got enough to eat, greatly increasing the distance that would need to be covered. Small parties were much more efficient all around, although predators, like large felids (Acinonyx, Panthera, Dinofelis, Homotherium, and Megantereon), were more likely to attack small groups. However, their climbing abilities helped keep them out of danger, and the females in particular never strayed too far in the open.

    Lucy did not take too long to find a tree laden with fruit, a particularly good one from her point of view since it was too small to support the weight of the heavier males and she was unlikely to be disturbed. Climbing up, she scanned the canopy to find the branches with the ripest fruit. Settling on a branch, she began plucking fruits with both hands and shoving them rapidly into her mouth. She sucked out the juice and pulp, wedging the fruit against her teeth, before spitting a soggy wad of skins onto the ground. Grunting in satisfaction, her strong front teeth bite through the thick skin of another fruit, and she ground the pulp and seeds with her powerful molars. While her front teeth were rather like a chimpanzee's, albeit narrower, her back teeth were very much larger and extremely well-suited to tackling the thick skins and seeds of a fruit, as well as the soft and juicy flesh. Indeed, Australopithecus afarensis teeth herald the beginning of a trend for increasing tooth size that would reach its peak, 2 million years in the future, in some of its descended relatives, which would be some of the most specialized hominids ever to evolve.

Having exhausted the supply of ripe fruits, Lucy climbed down, heading for some low-growing bushes that looked promising. Spotting a movement behind her, she turned and caught sight of more members of her troop arriving in the clearing, among them her son. She made a pant-like sound in greeting and he joined her. He had just turned 6 years old and Lucy saw him less often nowadays; for the past year he had been able to fend entirely for himself, and this newfound independence from his mother's milk meant he had been able to seek out other Australopithecus for company, particularly the adult males who hold a particular fascination for him.

Grunting contentedly, mother and son picked berries from the bushes. The infant, however, was restless and fidgety. Frustrated, her mother dumped her small, squirming body onto the young male, who began to groom his sister, seemingly content to act as a babysitter. The bond between these two would continue to develop as the baby grew, and she would always be able to look to her older brother for protection against aggression from either predators or other members of the troop, something she would need sooner than either of them realized.

Losing interest in food, the young Australopithecus pair began some rough-and-tumble games. Enjoying herself immensely, the infant pulled her lips over her teeth and pants at her brother, making a "play face." The peace did not last long, however. A distant screeching carried on the wind, and Lucy stopped in mid-chew, staring keenly in the direction of the calls. The other Australopithecus feeding nearby also stopped, and the males began to bristle, the hair on their backs and shoulders standing on end. Another scream cut across the clearing, closer now, and the infant began to whimper. Her brother patted her absentmindedly, for he was concentrating hard on the noise, trying to work out where it was coming from. Their mother, tense and agitated, took the infant back, pulling her close to her chest, and the infant began to suckle for comfort.

The young male became increasingly excited and, spotting a party of adult males on the far side of the clearing, bounds over to them, falling over himself in his eagerness to join them. Two of the highest-ranking males were present in the party, and with the alpha male now dead they were competing for his position. Among the privileges this brings were mating opportunities. The lead male had fathered most of the infants in the troop, including the infant female who had fallen asleep, her mother's teat still in her mouth.

While the cautious contender continued to scan the bushes, the other males screamed and chattered wildly to each other, attempting to gauge his reaction, waiting to see if he would lead them off towards the source of the noise. Unlike this morning, these distant shouts did not herald a predator attack, but the imminent arrival of another Australopithecus afarensis troop, intruders attracted by the high-quality foods available in this territory. The threat of violence was real, for both males and females, and the individuals were right to show fear. The intruder troop was larger and had more adult males. They had already beaten one of the rival males to death and attacked a pregnant female in the last month. The intruders seemed intent on eradicating this troop and taking over their range for themselves.

The other contender for the alpha position,

youthful and less experienced, lacked the guarded aggression of the cautious male and found it hard to keep control. As the males worked themselves up into more frenzied bouts of screaming, he grabbed a fallen branch and began flailing it wildly around his head. The females watched, seemingly impressed with this show of strength, while the brash contender whooped in excitement. Bolder now, he swung his newfound weapon at the other male, forcing him to jump for safety. The male shouted and charges, but it was too late. He had shown his opponent that he was vulnerable.

Shaken, but in control, the cautious older male left the clearing, followed by the young contender and two of the other adults. The adolescent male also tagged along. He was still excited, and one of the older males threatened him; this was no time for juvenile high spirits. The other males were tense and edgy.

Chastened by the threat and picking up on their moods, the adolescent fell silent. The cautious contender led them in single file, at a fast, determined pace. Occasionally, he left his position at the head of the patrol, climbing high in the trees to check what lay ahead. Shouts were still heard periodically from deep in the woodland, and the males paused briefly, before replying with shouts of their own. The adolescent male fell behind a little, his legs not strong enough to keep up with the pace. They had traveled at least over a mile already, but their range was broad, and they had a way to go. There was still no sign of the intruders, but the shouts were getting close.

The pace was stepped up once more, and the adolescent male was beginning to be left behind all together, when the cautious male suddenly stopped, bringing the patrol to an abrupt halt. The calls were directly in front of them. He let out a terrible scream of his own, then silence fell. The stillness was unnerving and the tension among the males was palpable, each one rigid with concentration. They were close to the boundary of their range, and their leader was wary. The contender, however, still beyond his earlier victory, was wild and unruly, flinging himself into trees, screaming hysterically. Unable to stay put any longer, he set off towards a large clearing just visible through the trees. It was the shortest route to the boundary, but it left them completely exposed. Not only was he being foolhardy, but he had also made another direct challenge to his opponent, who had been making decisions about the route, and he could not tolerate this.

    Their confrontation was brief but loud. Face to face, the younger male realized how powerful the older male was, and he suddenly lacked the confidence to continue his challenge. What is more, the other males showed little sign of supporting him in his efforts. The brash contender dropped to the ground, pant-grunting and holding out his hand as a signal of submission. Acknowledging this gesture with a hand on his head, the cautious male set off in his chosen direction, leading them down towards a fast-flowing river. The rest of the males followed without hesitation, all except the young male, who remained in the clearing. The cautious male turned to check that he was following. Instead of averting his gaze from his victor, as protocol dictated, the brash, young male met his eye. His baleful glare was not a good sign. This morning's confrontation would not be the last.

Like chimpanzees, Australopithecus afarensis males sometimes formed strong relationships with each other and spent most of their time together. Females formed looser, more casual bonds with males. When females were fertile, temporary courtships between males and females formed and they may disappear together for days on end but, as soon as mating was over, the band broke down and they went their separate ways.

At this stage of hominin evolution, there was no need for exclusive pair bonds or any division of labor between the sexes. For the moment, males and females were not dependent on each other to cope with life's daily problems. However, the shift to strict bipedalism was beginning to generate evolutionary pressures that would change this state of affairs dramatically, the results of which would be left right down to the future Late Quaternary.

While males and females led independent lives,

males were heavily dependent on each other to achieve dominance. They spent a great deal of time grooming, forging bonds, and building alliances. An alpha male, in particular, would take the trouble to ensure that the other males support him. He would be established and have many allies. An alpha male would be in peak physical condition and an impressive combatant. Another male would not stand a chance of defeating him, unless the dominant should suddenly lose power, either through injury or illness or by losing the support of others. Otherwise, the dominant male would retain this position until he died. The new leader may have to deal with dissent in the ranks; however, this was the least of his problems at the moment. Across the river, someone was watching them…

    The intruder male watched as the patrol moved towards him on the far bank. Sensing his group mates pushing up behind him, he threatened them with a soft coughing bark and they moved back. He shifted position for a better look and the branches around him shook. Back on the other side of the river, the movement caught the dominant male's eye. Realizing he had been spotted, the intruder male emerged from the foliage, staring at the dominant, challenging him. The males in the patrol began to sway from side to side, agitated, and once again their hair began to stand on end. Behind the intruder, there was further movement as the remaining members of his party emerge from the bushes, and the two groups face off across the water. The skirmish erupted in a frenzy of screams and hoots. The males in the patrol became wildly aggressive, jumping from the trees to the ground and climbing back again, tearing off small branches and hurling them across the river. The intruders responded in kind, heaving rocks, many of which fell short of their target, sending fountains of water into the air, adding to the chaos. The brash contender, still harboring aggression and resentment over his failed dominance bid, released all his pent-up hostility in a magnificent threat display, leaping around on a low-hanging branch jutting out over the water. Some of the males on the other side began to display similarly, but neither party seemed to want to take things further. The sides were too evenly matched to risk real violence; their displays were merely an effort to reinforce the boundaries between their ranges.

Gradually, the young contender's display began to wind down. He leaped onto the bank and landed close to the adolescent male, who greeted him with a barrage of frantic hoots and barks. Listening to their frenzied scream, it was hard to see these creatures as anything other than apes. But as the counter came to an end, they stand upright and walked away on their two hind legs.

Back in the clearing, the females had been feeding ever since the temporary departure of the males. So intent was they on eating their fill that it seemed no time at all before a hooting from beyond the clearing heralds the males' return. The adolescent male, now more boisterous than ever following his sojourn in the adult world, ran to Lucy and pulled his sister from her arms. He began to chase her around the clearing, working off the rest of his nervous energy, glad to be back among his family. His mother, meanwhile, greeted the lead male, and he offered his shoulder for grooming. Smacking her lips, Lucy settled down beside him and was soon engrossed in her efforts; until that was, her son got too rough with his infant sister, who began to scream in fear. Breaking off from grooming, Lucy was forced to rescue her youngest offspring, threatening her overexcited son as she did so. With her daughter hanging from her side, Lucy retreated behind some bushes where shielded from the rest of the troop, she attempted to calm the infant.

    Only the brash contender noticed her leave and, surreptitiously checking on the rest of the troop, he followed her. As he left his position, the calls of the rival troop could be heard once again; this morning's boundary dispute did not decide things after all. The dominant male, hair on end, ran from the clearing, Lucy's son and the rest of the adult males closed behind. The young contender watched them pass from his vantage point behind the trees, but he did not join them. Instead, his gaze returned to Lucy who, oblivious to his presence, played with and groomed her daughter. The male approached the pair. His intent was sexual, although Lucy was not yet ready to mate again. She was still breastfeeding and this inhibited the release of eggs from her ovaries, so for the moment, she was infertile. However, if the young contender could engage her in grooming now, forming a bond between them, she may choose to mate with him when the time was right. The female was still unaware of the male's presence, however, and was lazily picking berries, watching as her infant discovered the joy of playing with sticks. The male reached out and touched her on the shoulder as a prelude to grooming. Alarmed, Lucy spun around, and, surprised to find herself alone with him, she barked in fear.

The male, in his turn, was unsettled by Lucy's

reaction. Still tense from the morning's frustrating interactions with his cautious opponent, he seized her roughly and began to grapple with her, his intention merely to force Lucy to groom, but she resisted. His body formed a barrier between the mother and her infant, and the little one begins to whimper. Hearing her baby cry, Lucy tried to push past the male, her arms flailing. Misconstruing her action as an attack, the male snarled and launched himself at her, landing a punishing blow on her back. What was originally intended as a harmless exercise in bridge-building has mutated into an all-out fight.

The other adult males close enough to hear Lucy's screams, rushed back, her son leading the way. He hurled into the clearing to find his mother facing off against the young male. Moving between them, he tried to defend her. The male moved forward menacingly, but the arrival of the other males stopped him, and he backed off. Lucy immediately scooped her infant from the ground, once again having to comfort her terrified offspring. But the aggressive male had not finished yet. Facing the other males, he drew his lips over his teeth in a grimace, half-fear, and half-threat. Unsure what he was up to, the other males watched him closely. Suddenly, before anyone had time to act, he leaped at Lucy, knocking her to the ground, snatching the infant from her, and bounding off into the undergrowth. In his highly charged, testosterone-fuelled state, the male was about to attempt one of the most shocking of all foul behaviors: infanticide.

Sometimes, certain animals attack and kill youngsters of their own species, helping to increase a male's chance of mating by bringing the mother back to a fertile state, by removing the inhibitory effect that (in a mammal's case) suckling has on egg production. Second, by killing another male's offspring, he reduced their reproductive success relative to his own, so that more of his genes were passed on to the next generation.

As the male tore through the undergrowth, Lucy and the other males gave chase, desperate to save the infant, heedless of the thorns that caught on their hair and skin. Their shouts and screams carried to the brash contender, who was losing ground as they moved into the dense woodland that ran along the river. Confronted by a tangle of thorns that he could not penetrate, the male was forced to turn and confront his pursuers. Dropping the infant, he displayed at them with ferocious snarls, while the infant, bruised, shaken but thankfully alive, crawled into the undergrowth to safety. The male stepped up his display, but it was the lead male who attacked first, throwing himself at the young contender and forcing him to the ground. The younger male fought back, pounding on the older male with his long forearms, screaming at the top of his voice. Lucy launched herself at him as well, screaming in both fear and anger.

Amid all this chaos and violence, the rest of the Australopithecus in the troop gradually became aware of more shouting coming from just beyond the clearing. At first, they did not realize what was happening, but as several strange Australopithecus males appeared before them, they understood that the rival troop had returned. Outnumbered by the intruders, the resident troop members shouted in fear, realizing their advantage, the rival troop attacked….

Brotherhood

Period: Early Quaternary (Early Pleistocene, Middle Gelasian)

Date: 2,000,000 BC

Location: Eastern Africa

The reed bank rippled and swayed in a light breeze and dragonflies flitted across the water of a lake, flashing blue and silver. The ferocious heat of midday had died away but it was still hot and very humid. On the horizon, clouds were building. Soon the rains would come and the long dry season would be at an end. At the edge of the cape thatching reeds, where the receding waters had left a bank of dried out mud, a 4-foot-and-6-inch tall, sturdy male great ape crouched, digging into the mud with a stick. Reaching down into the hole he had made, he pulled up the muddy root of a carex and, after briefly rinsing it in the water, popped it into his mouth and began chewing. Large, strong chewing muscles bulged in his jaw and cheeks as he ground away at his food, steadily doggedly, the mouthful of food taking almost half a minute to consume. His odd, wide face appears impassive, but he was keeping a close eye on the rest of his troop, who were strung out along the shoreline.

Turning his head, he also monitored the few troop members that were digging for food in the shade of the trees that fringed the lake. They also used sticks to dig but, instead of carex roots, blubs and tubers were their reward. Too far from the water to make washing an option, they rubbed their food against their hairy arms to dislodge dirt and grit before eating it. The only sounds to be heard were the contented grunts they made at each other, and the sound of relentless chewing, as they pulverized food between their relatively large back molar teeth. These Creatures belonged to a genus known as Paranthropus, this species being Paranthropus boisei, and they were Australopithecines, the same sub-tribe as Australopithecus.

To the amazement of the Paranthropus, the peace of their feeding session was broken as another bipedal great ape (a male) entered the glade, running at full tilt. Like the Paranthropus, his kind was short, on average standing no more than 4 feet and 3 inches tall (but sometimes a little taller), with long arms and short legs, but there the resemblance ends.

Unlike them, he was lightly built, and his hair was rather sparse, revealing the dark skin beneath. He was also more finely featured, with a sloping forehead and a less protruding face; he lacked the huge jaws and massive facial muscles that gave the others their strange wide faces. His teeth were also considerably smaller and narrower than that of any species of Paranthropus (a hard, mostly fibrous diet was not something his kin was designed to cope with) and his brain was larger. This was an early variety of humans (the genus being Homo), of the species Homo habilis, and he and his kind had developed skills unseen on this planet until now.

    As he vanished through the trees on the other side of the clearing, the reason for his high-speed exit appeared. They were also great apes, but slightly taller than the Homo habilis, and their long-legged build was a bit more modern than his; all three of the pursuers ran with the same easy, fluid gait. Facially too, they were distinct from Homo habilis, with a rather flat face that brought to mind a much less extreme version of a Paranthropus boisei. They also had much larger teeth and, intriguingly, their brains were larger as well. They belonged to a more primitive-looking species of human called Homo rudolfensis. They shouted and gestured at each other as they ran, clearly trying to decide which route they should take. Then they too were gone, their shouts fading in the distance.

Over the past 1.5 million years, the great apes, already bushy to begin with back in the Neogene, had blossomed into an even more fantastically rich and diverse array, there were now as many as half a dozen species in existence, each with their own strategy for survival.

The answer to what caused this remarkable explosion of great ape species laid in the gradual drying of the East African climate that began back in the Middle Neogene. Apparently through the decline of wooded savannah and seasons becoming more extreme, new species appeared and went extinct, leading to what had been called a "turnover pulse" of evolutionary change.

In East Africa, 29 or so species of forest-dwelling bovids became extinct and were replaced by species that were specialized for living in open habitats. Species previously found only in northern regions of Africa (some from Europe) also began to appear in the southern portion of the continent as the planet cooled and temperate habitat types spread towards the equator.

By the Early Quaternary, a whole new suite of animals was then inhabiting East Africa, all of which were adapted to the new range of habitats and food sources that climate change had created, and the hominids were no exception. They too had diversified into several species, each with a unique array of adaptations that enabled them to meet the challenges that life now presented.

While the Paranthropus boisei was specialized to feed mainly on low-quality, abundant foods such as underground tubers and roots, the two early human species of the area (Homo habilis and Homo rudolfensis) had evolved a different strategy, greatly varying their diets with the seasons and aiming for the highest quality food sources they could find. Both of these strategies were highly effective, but only would prove to be resilient in the face of continued climatic change.

The defining mark of humans is increased brain size. Homo habilis had an average cranial capacity that ranged from 550 cm3 to 687 cm3, still a long way from the 1350 to 1450 cm³ range of modern humans, but a mark increased compared to Australopithecines. This increase in brain size reflected the expansion of one area in particular; compared to the Australopithecines, Homo habilis had much bigger frontal lobes, and the pattern of wrinkles and grooves on them was the same as in future human species. These parts of the brain were associated with higher thought processes, such as problem-solving and planning, making Homo habilis capable of the kinds of abstract thought far surpassing the abilities of any Australopithecine. Interestingly, large brains did not characterize all Homo habilis. There were some with an estimated cranial capacity of only 510 cm³, around the same size as that of a Paranthropus boisei. However, a Homo habilis brain also had a different structure from that of the Australopithecines and this change in brain organization may also be accounted for the Homo habilis's increased mental capacities, rather than the increase in size alone.

Regardless of what actually explains Homo habilis's greater mental skills, there was no doubt that they placed them to good use in their ceaseless search for food, and one kind of food in particular: meat. This was especially so now, in the dry season, when fruits and other plant-based foods were in short supply. At times like these, Homo habilis turned increasingly to meat scavenged from predator kills to supplement its diet, and a larger brain was a definite asset for many reasons.

In the first place, finding fresh kills required Homo habilis to be aware of the clues available in its environment, since carcasses were widely scattered and so random searching was highly inefficient. By taking their cue from other scavengers, such as the white-backed vultures, Homo habilis saved valuable search time. Second, once a carcass was located, a Homo habilis's ingenuity helped it to protect its precious find against competing scavengers, such as hyenas and canids, and also the predators themselves.

Lastly, and most importantly, Homo habilis's larger brain meant that it possessed a skill that enabled it to exploit a carcass in ways far beyond the means of other scavengers: something that represented a major breakthrough in the hominid line, marking them out from the rest of the animal kingdom; something that the young Homo habilis male had in mind when he paused to grab a stone on the way to his first decent meal for a week.

The rest of the Homo habilis party followed the young male, and soon they were perched high on a rocky outcrop overlooking the dry, scrubby plains. Below them lay the carcass of a recently killed bovid, specifically a young common eland, on which the White-backed vultures were feasting. The Homo habilis troop scrambled down, shouting as they went to drive off the vultures, their excitement at finding food overcoming caution. The young male, however, hung back. The kill was very fresh and he knew the predator responsible may still be close by. However, it was not necessarily a predator that he needed to watch out for at the moment.

The other Homo habilis reached the carcass.

The dominant male was there first and dropped to his knees to inspect how much meat remained. As he did so, a stinging blow to the head knocked him forwards, and he found himself sprawled across the fly-covered eland body.

A large stone, now lying harmlessly in the dust, seemed to have been responsible. Another stone, and then another, confirmed this to be the case. Fending off the rain of stones, the troop turned to see a group of Homo rudolfensis silhouetted against the sky, high on the outcrop that the Homo habilis troop had just left. Although these omnivores ate primarily on plant matter, both Homo habilis and Homo rudolfensis scavenged for red meat when times were hard, and the two species often found themselves in direct competition with one another.

Partly this was because they could only feed on fairly fresh carcasses. Rotting meat made them ill, and they were not attracted to anything that had been dead for any more than a day or so. The Homo rudolfensis also saw the circling accipterids and realized there was fresh meat for the taking.

The Homo rudolfensis descended to the plain. The dominant Homo habilis male got back on his feet and ran forward to meet them, followed by the other memberrs of his party. The two primitive human species displayed each trying to get the other to submit and leave, without actually coming to blows. The display intensified but still, no blows were struck.

Fighting used up expensive energy and was potentially dangerous (a wounded Homo habilis or Homo rudolfensis found it hard to keep up with the rest of the troop and could not climb trees or cliffs to avoid danger) so individuals, mostly males, tried to avoid actual fighting whenever possible. Today, however, the rewards were too great for either side to give in.

The aggression escalated and the dominant male Homo habilis launched an attack on one of the male Homo rudolfensis. They wrestled and gouged at each other's faces, aiming for the eyes, striking blows wherever they could. The Homo rudolfensis male struggled underneath the smaller but tenacious Homo habilis and managed to lever him off by pushing his feet against the Homo habilis's chest. The Homo rudolfensis male got ready to launch himself at the Homo habilis, who now lay sprawled on the ground, but something caught the eye of the Homo rudolfensis and he looked up. Horrified, he shouted to the rest of his troop who, without hesitation, began sprinting back to the rocks. The dominant Homo habilis shouted in victory, believing he was responsible for the retreat. Turning to the rest of his troop, he saw they were also fleeing in the same direction as the Homo rudolfensis. He barely had time to register a feeling of surprise before two Panthera crassidens leaped onto him from behind. Unnoticed by the brawling great apes, they had returned to their eland kill. The Homo habilis's skull was smashed by huge canine teeth, killing him instantly, and the felids began to feed, both on the ape and the antelope. As the remaining Homo habilis watched from the safety of the rocks, it became apparent just how small and vulnerable these simians were. Living by their wits alone in a habitat full of large, powerful predators was a risky game and Homo habilis life was harsh and often short. Many individuals died at around the age of 15 years, still very young indeed for Homo habilis. The causes of their deaths were varied (usually from attacks by felids and crocodiles), and it showed that life was far from easy for them.

The Paranthropus boisei, on the other hand, did not face such hardships. Their diet may be bland and monotonous, but it was not dangerous, and specializing primarily in food that was so widely distributed and plentiful meant that their populations could grow much more rapidly than those of Homo habilis and Homo rudolfensis. With a little ingenuity, however,

Paranthropus could introduce some variety into its diet, even during the dry season.

Having left the dominant male and the dominant female behind, the new female Paranthropus moved off to feed. Hunger and her fear of the dominant female had got the better of her and she decided to stay away from the troop for the rest of the day. Clutching a digging stick, she scanned the ground in front of her, looking for the small, dry shoots that indicated where there were corms to be found, but then she spied something far more exciting.

Approaching a termite mound, the female circled it, looking for the best way in. Taking the digging stick in her fist, she poked at the hard dirt crust until pieces began to break away. As she did so, the first termites began to emerge. Quickly, the female scooped them up in her hand and shoved them into her mouth, crunching them down before they had time to bite back in resistance. The female Paranthropus managed a second handful and a third, but then the soldier termites came teeming out in defense of their nest, exuding a foul sticky substance as they did so. Wiping her hand across her face, the ape backed off, satisfied with her termite snack.

Glancing up, she was startled to see that the dominant male had followed her. Coming close, he poked the second hole in the mound, on the opposite side to the first, and was able to grab his own termite snack before being defeated by the remorseless soldiers.

Discarding the stick, he presented his shoulder to the female, and she began to groom, nervously checking for the dominant female, but she was nowhere in sight. The male began his mating courtship in earnest, lip-smacking and grooming her in return, and eventually the female relaxed. As the last rays of hazy sunshine broke through a storm-darkened sky, the dominant male Paranthropus achieved his goal and, through the act of mating, added a new female to his harem.

Returning with his new mate to the rest of the

troop, the dominant male began to climb up into a tree to rest for the night. The new female attempted to follow, but a coughing threat from the dominant female led her to abandon her attempts to share the male's night nest. Instead, she retreated to her own tree. She made a desultory nest for herself and, despondent, settled down for the night as the first few dropped of rain began to fall.

    The dominant male, meanwhile, decided that making a nest was too much effort and made his way over to the dominant female, intending to share hers. She refused to make way for him, and her dominant status, plus her animosity to the new female Paranthropus, meant the male was reluctant to push her any further. Moving slowly through the massive branches, he approached another female and another, but their nests were small and the male could not fit.

Finally, as the rain began to fall in earnest, the male forced himself into the nest of the lowest ranking female, squashing her and her infant in the process. Despite these efforts, most of him remained hanging out of the nest but, seemingly oblivious to his ludicrous situation, the male settled down for the night.

The next morning, the Paranthropus awoke to a water-drenched world. The grassy clearing was festooned with rain-sparkled spider webs and the Paranthropus's digging holes formed small pools of water from which birds flitted to and fro, bathing and drinking. The dominant male Paranthropus climbed down slowly and shook himself. His night was long, wet, and miserable. In need of some grooming to help him dry off and warm up, he moved over to the part of the tree where the dominant female slept.

Climbing back up, he poked at the dark, hairy body that still lay asleep. It stirred and the dominant female's face appeared over the edge of the nest. The male grunted in greeting and gave a start as another hairy Paranthropus head appeared. The new female grunted sleepily at the male, and then at the dominant female. The ferocity of the storm disintegrated the new female's paltry effort of a nest and forced her to seek shelter in a new tree. She was apprehensive when she heard the soft grunts of the dominant female inviting her to approach but, not wanting to spend a night in the rain, she took up the offer.

Now it seemed that her problems were over; having gained the dominant female's acceptance, her new life in a new troop could begin. It was the dominant male who now had to appease the dominant female, and, smacking his lips together softly, he began to groom her, his cold, damp fingers searching through her thick, coarse hair.

For Paranthropus boisei, the trials of life arose mainly from the complexities of living in a tight-knit social group: how to form relationships (friendly or intimately), find mates, raise offspring, and, in the case of the dominant male, kept the peace between the females. Their superb adaptations to their diet, and the ease with which they could find the food they needed, meant that Paranthropus was rarely filled with the kinds of trauma that dogged the Homo habilis. Life for the most part remained the same, day in and day out, and as a result, Paranthropus was not especially innovative. Unlike Homo habilis, who needed to be alive to all possible opportunities that came its way, Paranthropus could stick to what it knew best. Such specialization was not to be confused with dullness or a lack of intelligence.

Indeed, Paranthropus brain size would actually increase over time; later specimens would have brains so much larger than the earlier individuals that males would no longer display sagittal crests (their skull alone will be big enough to allow for the attachment of the temporalis muscles). However, such a highly specialized diet and lifestyle meant that Paranthropus had less ability to cope with changes to its environment. It had gone too far down one evolutionary pathway to change its ways with any ease. This would prove to be its downfall.

Over the next few hundreds of thousands of years, the dramatic climate changes that shaped Paranthropus's environment were set to continue. The climate would become increasingly erratic; swinging from hot to cold in a series of ice ages and the landscape would continue to undergo tremendous geological change: change that would mean extinction for many species.

Sadly, Paranthropus was one of them. Despite its behavioral complexity and its feeding specializations, the Paranthropus genus would die out in 1.2 million BC. One reason for this eventual extinction would be the change of climate reducing the number of reed beds available so that their dry season root-eating strategy would fail them. Another reason would be competing with the other great ape species for high-quality vegetable foods increased, so that Paranthropus would have to rely more heavily on their poor-quality alternatives, even during the wet season, and in the end, they simply would run out of time, unable to spend enough hours each day feeding to get the energy they needed.

Whatever the reason, Paranthropus would eventually become fossilized. But this did not mean that they had no part to play in the hominin story. For without such Paranthropus species as Paranthropus boisei, the early species of humans may never have evolved. The presence of Paranthropus during that time meant that the ancestral humans were denied the opportunity to occupy the same primary vegetarian niche, and were forced to evolve alternative means of coping with the new habitats of East Africa (an opportunistic, "try-anything" strategy).

Without Paranthropus providing them with indirect competition for resources, they may never have evolved the technological skills and social patterns that will sharpen the pattern of human evolution. They may not be the direct ancestors of all human species in general, but Paranthropus was still one of the reasons that humans would eventually succeed in the future.

High on the rocks, the remaining Homo habilis continued to watch the two Panthera crassidens fed until it was too dark to see. Unsure what to do without the dominant male, they settled down to sleep where they were. They were safe on the cliffs and huddled close together for warmth. With empty stomachs, sleep took a long time to come. The young male, however, was more hopeful than the rest. Less distracted by the dominant male's demise than the others, he had noticed the huge storm clouds building up on the horizon and he knew that rain was on its way, bringing the dry season to an end.

The storm lasted all night and, when morning comes, the Homo habilis sat on the cliffs, small and bedraggled, gazing down longingly at the common eland carcass. The felids were still there; one was asleep while the other gnawed lazily on the remains of the deceased dominant male Homo hablis.

While the rain signaled that a new flush of growth would soon begin, the Homo habilis needed food right now, and the dead eland remained their best bet, even though most of the meat had now gone, picked clean by both the Panthera crassidens and the white-backed vultures.

The lack of meat was not a problem because the Homo habilis was after something else: the sticky marrow inside the bones. Its high-fat content meant it satisfied their hunger quickly and it also served a physiological need, because fat was needed to digest protein.

Without a good supply of fat in their diet, the Homo habilis would not be able to eat even the small amount of meat remaining on the carcass, as they would have to use their own body fat reserves to digest it (defeating the object of eating the meat in the first place). Extracting the fatty marrow was one of the keys to their survival, but it was not easy to get at and few other animals were able to exploit it. But Homo habilis had a secret weapon.

Squatting on the cliffs of the outcrop, the young male picked up the stone he brought with him yesterday. Selecting another stone from those scattered around him, he tested its weight in his hand and adjusted his grip.

Making sure he was holding both stones firmly, the male brought his arm down, striking one stone against the other. A large flake chipped off the stone that was struck and fell to the ground. Adjusting his grip again, the males struck another flake from the core stone, and then another. The male sat back and inspected the object he had just made; it was a stone chopper, crude but highly effective, it was a tool that he could use to smash open the bones of the eland and extract the tasty, rich marrow.

The flakes, with their sharp edges, can cut through flesh and sinew easily, allowing Homo habilis to dismember a carcass, removing the meat and marrow-bearing bones, and transporting them somewhere safe for feeding.

The precarious, scavenging lifestyle of Homo habilis had bred into them a remarkable ingenuity and resourcefulness which, combined with their advanced reasoning skills, gave them the power to overcome their inherent disadvantages as mostly carnivorous omnivores. Lacking the same sharp teeth and claws that allowed certain other animals to feed on prey, Homo habilis had solved the problem by making tools that would do the job for them.

The significance of this development could not be underestimated. While certain other apes, like Paranthropus, may use digging sticks, these did not require a very deep understanding of tool-making. It was easy to see how a leafy stick could be turned into something suitable for digging by stripping off the leaves and small stems. It was another matter entirely to understand that by striking a shapeless lump of rock at the right angle, and with the right force, a sharp-edged blade can be produced.

Chimpanzees have similar skills needed to fashion tools like those of humans such as Homo habilis, beyond the reach of many other apes.

As well as mental agility, the stone tools of Homo habilis also indicated increased levels of manual dexterity. The hand bones of Homo habilis had straight fingers with broad tips that were well-supplied with nerve endings, increasing their sensitivity. One of their thumb bones also had a broad head, compared to Australopithecus and other great apes. This was associated with the fine precision grip of modern humans, and Homo habilis was also able to grasp objects between the tips of the thumb and the other fingers (an attribute that was crucial for effective stone tool-making).

With this combination of dexterity and inventiveness, Homo habilis was able to produce a whole range of tools. As well as flakes and choppers, there were hammer stones, coreforms, and manuports. Homo habilis also further work flakes and cores after their initial manufacture to improve their effectiveness. Many flakes could be retouched along with one, and these were termed 'scrapers'. Choppers could also have flaked on both surfaces instead of just one, and these were known as 'bifacial' choppers.

In most cases, however, Homo habilis tools were crude and uneven, and their final shape seemed to have been determined by the initial shape of the rock, rather than an idea in the head of the toolmaker. Although they had made the massive conceptual leap needed to make tools, Homo habilis was not particularly skilled toolmakers. As a result, the different kinds of tools tend to grade into one another, rather than forming completely distinct types.

Homo habilis also made tools as and when they were needed, rather than using the same ones over and again. Their only signs of forwarding thinking were the manuports, which meant that Homo habilis carried certain kinds of rocks with them when out scavenging, perhaps because these were the most suitable for toolmaking. But regardless of their crudeness, the stone tools of early humans showed that they had reached a new level of intelligence and understanding. With this truly extraordinary invention, these toolmakers had set their generation off on a path that would culminate in such mindboggling achievements.

Grasping his freshly minted stone chopper, the young Homo habilis male rounded up the rest of the troop. They also had freshly made tools and were ready to eat. Peering down at the kill site, they see the two Panthera crassidens were still there. They had not moved off in search of fresh meat as the Homo habilis hoped. While most of the troop gazed down gloomily, the young male stared into space, apparently deep in thought.

A moment later, he jumped up and, grabbing another lump of rock, began to bang it against the stone chopper noisily. Deciding this was not quite noisy enough, he began to yell, and he also threw more rocks down towards the Felids for good measure. The other Homo habilis the idea and they joined in with the male's noisy display, which was clearly designed to scare the panthera away from the kill.

Still yelling, the young male then surprised

everybody by leaping off down the rocks, straight towards the Panthera crassidens. Less sure this time, the other Homo habilis followed suit, and the raucous mob descended on the carnivorans, which now retreated carrying away the body of the deceased dominant male Homo habilis with them, unsure of what to make of it all. With some relief, the male saw that his risky gambit had worked. Usually, only the larger Homo rudolfensis could use this tactic to see off large predators, but the sheer desperation of the Homo habilis to get at some food had brought success on this occasion.

Falling on the carcass, the Homo habilis began to saw away at the common eland's large back legs with their flakes and choppers. Severing them from the carcass, they retreated back up the rocks to safety; where they began smashing open the bones with hammer stones, picking out the marrow with the ends of their sharp flakes. It was rich, sticky, and delicious and the Homo habilis chattered and grunted happily as they ate their fill. The young male in particular was exhausted but content. His skills in driving off the panthera and ensuring a decent meal for the troop meant that most of the other Homo habilis in his troop was treating him with a new deference, particularly the females, who were keen to offer him the juiciest samples of marrow they could find.

If the young male could capitalize on this, his future as the new dominant male in the troop seemed assured.

Unlike many of Homo habilis's fellow apes, skill and ingenuity were more important attributes for gaining position than sheer brute strength. With the harsh dry season over, the life of this Homo habilis troop would be easier, and the young male was sure that his foraging skills would stand him in good stead during the coming months. The dangers and hardships of the past few months forgotten, the Homo habilis sucked contentedly on the marrow-filled bones and began to look forward to the lush possibilities of the wet season.

Homo habilis had taken the next step towards their faraway future descendants. Its adaptability and tool-using skills meant that it had begun to free itself from the rules governing almost all other animal species on Earth. It was beginning to gain control of its environment.

But despite these skills, Homo habilis would not be the exact species that dominated Africa over the millennia to come, for another human species would soon be emerging here on the savannahs of Africa; it would be a species that also made stone tools, but tools that were of extraordinary beauty and design, revealing a subtlety and skill far beyond the skills of Homo habilis. This species would have a larger brain than Homo habilis and possessed a more advanced array of mental skills. It would be a species that would look and act a bit more like its descendants. Its name would be Homo ergaster.

PRECURSORS

Dysfunctional Family

Period: Early Quaternary (Early Pleistocene, Early Calabrian)

Date: 1,500,000 BC

Location: Eastern Africa

It was the middle of the day. The sun blazed overhead in a clear sky. A vast dried mudflat stretched away as far as the eye could see, the air above it shimmering in the relentless heat and the cracked ground burning to the touch. But the male creature pacing on the baking surface did not notice the scorching temperature. He was walking steadily and evenly, unfeeling of any possible pain beneath his feet, focused on one thing only: his prey.

In the far distance, a lone 5-foot-tall blue wildebeest was moving slowly. It was old and tired, and nervous. Separated from its herd, it was exposed and in danger, but here on the baking mudflat it had found a respite from the heat. A small, muddy waterhole was offering the bovid a chance of recovery, but it could not linger there long, for it knew that it was being stalked. This was East Africa, and here the evolution of the hominin line had taken another extraordinary step.

More advanced humans had, at last, emerged on Earth. The wildebeest was being stalked by a human species known as Homo ergaster, and the difference between his appearance and that of the smaller Homo habilis that had already emerged before his kind (and still existed in that time) was immediately striking. Even the feet taking the Homo ergaster on his relentless pursuit were clearly more advanced in almost every detail: slim, long, toes reaching forward. The lower legs were long and slim too, supporting knees like future species, and long muscular thighs.

Perfected by evolution, after more than 2 million years of upright walking in other primates, these were the limbs of a tall, more advanced type of human. His hips were narrow, his stomach flat, taut and strong, his chest widening above it, breathing steadily, calmly, and not panting. His spine was thoroughly straight; his shoulders broad, and at his sides hung long slender arms with narrow grasping fingers. His skin was a dark olive color, smooth but with gentle downy hair on the limbs and chest. He stood 6 feet and 2 inches tall, with all the physical attributes needed for his environment: an African animal at home in his harsh domain.

Yet at the top of this very human body there rested a distinctly non-advanced human head. A heavy jaw, teeth bared into something far more threatening than a smile, a flattish nose, a low forehead masked by a mop of straggly hair, and a thick brow ridge that shielded two bloodshot eyes: the face of a typical ape, but an ape with new intelligence.

Now the Homo ergaster stopped still, comfortable with his distance from the ailing wildebeest, which, exhausted, collapsed resignedly into the mud. The hunter's eyes focused on it attentively. The hair on his head was greying, marking him out clearly as an older individual. Beside him were other similar Homo ergaster; together they made up a small party devoted to gathering food. They all stood in silence, watching the weakening wildebeest.

For Homo ergaster, feeding was rooted in opportunism and patience. They did not have the speed to be pursuit predators or the strength to ambush large prey.

Typically, they scavenged whatever meat they could find that had been leftover from the kill of something like a large felid, and as a troop, they were more than able to see off any vultures that may have congregated by a carcass. But when the opportunity presented itself, they would themselves take an old or ailing prey item. Their method was slow but effective: their bodies had evolved the limbs and stamina to walk and stalk their quarry into the ground. Searching for fresh meat could last a day or more, but the rewards were great.

The older male stood impassive, watching the prey. He was judging the moment to move closer. An older female stood near him. Clearly, past childbearing age, she too could help with the slow and steady hunt. Beside them a younger male, who shall be called Hoovu, appeared more on edge, eyes darting from view to view. A successful kill was a rare event for this species, and he was excited.

Suddenly and without prompting he took off, running past the older male and the rest of the group. A second young male followed him, and the two rushed towards the wildebeest, brandishing wooden stabbing implements. The two Homo ergaster shouted and yelled, disturbing the quiet of the landscape as they bore down on the prostrate even-toed ungulate in front of them, confident of their attack. But gradually their calls fell away as the mud around the waterhole started to slow them down. Then, extraordinarily, there was a twitch of life from the blue wildebeest. Roused to a final effort, it took a huge breath and, as the two humans continued to struggle through the mud towards it, it rose to its feet and staggered on. Suddenly it found firmer ground and, with the last gasp, it was free and on its way to join the rest of its herd.

The opportunity for a kill was lost. Hoovu and his friend knew they had made fools of themselves and, as they extracted themselves from the mud and dust and slope back towards the group, the old female gesticulated, and barked strange nasal sounds at them.

If she had words to describe her feelings, she would call them impatient fools, but this concept of language was not that subtle. For the old male, however, their spontaneous interruption of his killing technique had passed without effect. He still clearly knew best and was still dominant.

Impassively, he turned towards the direction of the blue wildebeest herd, scratching his face with a conical Nile crocodile tooth that he has been holding. It was a trophy of a previous encounter, and he had learned that he was seen as being special for having gained it. His knowledge of the other animals they hunted, the techniques he used, the skills of the kill, they had all taken many years to perfect and his wisdom still counted. Meanwhile, there was that wildebeest to catch. The relentless stalking began again.

The key to Homo ergaster's survival was a set of remarkable evolutionary adaptations to the environment into which it emerged. For 2-3 million years after upright walking began among the earliest human ancestors, the world had been going through a relentless phase of gradual cooling and, as the icecaps at the poles steadily locked up more and more water in their frozen masses, so the world became drier too.

By 2 million BC, the dominant environment in East Africa was open, dry savannah, a precursor to the vast familiar grasslands of the not-too-distant future. As the great apes began to colonize this landscape, the evolutionary pressure on them produced physical changes that had become the hallmarks of all humans. And the driver of this change was the sun.

At noon on this early savannah the temperature could be 35 degrees Celsius in the shade (but here on the flat there was no shade). The Homo ergaster was the only mammal hunters around at this hour. Other predatory mammals such as the felids, and even the hyenas, lay panting and dozing while the sun was at its height, but these great apes could seize this hunting opportunity from the competition. The tall, thin skeleton of Homo ergaster was a response to the hot climate, maximizing the surface area of the body from which to lose heat, and had evolved over hundreds of thousands of years to form this new heat-efficient species.

But Homo ergaster's evolutionary fitness did not end with the long and tall frame. Its thin upright physique meant that no longer did it need extra fur covering its body to protect areas directly exposed to the sun. Instead, it had a large area of smooth skin, and that enabled it to lose heat by a newly evolved, but highly effective mechanism. Homo ergaster cooled down by sweating, far more efficient than the panting used by many other fur-covered mammals. And it had another new, but familiar, human feature. It was the first great ape to have a distinctive nose. All previous great ape ancestors had nostrils that lay sunken into the surface of the face, but Homo ergaster's nose stuck out, allowing more air to be cooled and moistened by the lining of the nostrils before reaching the lungs, and at the same time allowing more moisture to be retained in the body on breathing out. The sweat glands and nose gave Homo ergaster the edge in the hot African day. Their lanky skeleton provided long powerful limbs, very strong, and ideally suited to running, while their new efficient heat management meant they could run, walk, and keep walking for a long time.

    The hunting band of Homo ergaster walked steadily on, untiring, while in the far distance the image of a herd of moving ungulates rippled in and out of focus through the heat-haze. It was impossible to see exactly what they were. The older male Homo ergaster paused and looked down at a series of regular marks on the ground. They were hoof prints and, tracing one with his index finger, he looked from them to the distant herd, making the connection. This distant herd must be made of giraffes. It may seem a simple act of observation, but in that single moment, the older Homo ergaster male revealed the secret of what really marked his kin out as a unique species of this time. It was not the remarkably advanced human body, but the thing that resided inside the non-human great ape-like head. For, at a volume of about 1000 cubic centimeters, Homo ergaster's brain was half as big again as the smartest of its predecessors, and almost within the limits of modern human variation. This new brain capacity had brought even greater powers of thought into the everyday life of this species.

All animals had some understanding of their environments. A 5-month-old swallow was instinctively able to negotiate a 6,000-mile migration from Western Europe to southern Africa without ever having done the journey before. An old matriarch elephant could remember where, in her herd's vast territory, to go for water at a certain time of year.

Primitive and earlier human species such as Homo habilis and Homo rudolfensis had already learned to associate different signs in their environment, such as the wheeling of vultures in the sky as a sign of a dead carcass. But Homo ergaster had taken that further, making complex deductions about apparently unrelated events going on around them. They could look at marks in the sand and, never having seen them before, could tell at once what they were, and what they were likely to relate to.

To a canid, a felid, or even to a baboon, hoofmarks such as these were no more than just that: random marks. Only Homo ergaster and its later relatives, of all the animals on Earth, could see them for what they were: footprints, made by another animal that was likely either to be a meal for them or to be hostile towards them.

Silently, the old male turned away and led the party on. Now he could make sense of the blurry movement shimmering in the distance, he knew it was of no interest to him. He knew his quarry, and he was gaining on it. But Hoovu stood a little way apart from the group, looking at some other markings. He too was trying to learn to make links. Suddenly the older male barked at him, clearly angry that this distraction was slowing them down. He did not want this kill to be spoiled for a second time.

The improved Homo ergaster ability to make deductions about the world they lived in, and applied them to totally unfamiliar things around them, had given them much greater adaptability than the earlier hominins. It was a milestone on the journey that would lead the later human species to be able to understand their environment.

Once again the hunting party had come to a stop. Clutched in the old male Homo ergaster's hand was an object that was the tangible proof of his abilities. At some time in the last half-million years, his kind had made a remarkable technological breakthrough. He held a beautifully symmetrical, teardrop-shaped stone hand ax, and its beveled edges glint with beauty and menace in the bright African sun. It was a stone tool that was the product of a thoughtful mind (a mind that could identify a nodule of stone, could hold a plan of how to turn it into a "knife," could carefully select the point at which to strike it with another rock, and then consistently work the flakes as they broke away, gradually refining the stone, first one side then the other, to a razor-sharp elegantly symmetrical blade).   Powerful and adaptable as a tool, it could even be touched up or sharpened; the creation of this hand ax showed planning and vision. They were of an order of sophistication greater than that of the bifacial choppers made by Homo habilis. During that time, it was the most complicated thing ever crafted on the planet.

The older male was trying to see track marks on the ground, but without success, and his anxiety was beginning to show.

Meanwhile, Hoovu was again becoming distracted. He was using his own hand ax to lever some bark away from a dried-up bush. Suddenly there was a flurry.

The same blue wildebeest from before was spotted, again on the ground, now too weak to follow the rest of its herd. The moment had arrived. Hoovu trotted up to the rest of the group again, and they all stood, tense and expectant, awaiting the command. But this time there was no dissent or premature action. The older male waited to be certain the wildebeest was really laboring again. Then it was happening. He made a barely perceptible signal. This was a process they had been through before and, once launched at the stricken prey, they would make a formidable team of killers.

Not crying out this time, they simply ran, their long legs moving almost effortlessly as they bore down on their prey, and the wildebeest had neither energy nor time to escape. As they reached it, the first-hand ax thudded down into the antelope's neck; then a second. There was no real defense, and the herbivore was soon dead.

Here the distinction between Homo ergaster and certain future human species was starkly drawn. The small group immediately began to satisfy their hunger on the carcass: tearing at the flesh and sinews with their powerful jaws and teeth, slashing more skin from the wildebeest's body with tier sharp hand axes, lifting their heads only briefly and sharply to watch and listen for approaching rival predators before bending again to the feast. Out on the baking mudflat, a fast, silent, and bloody meal was being taken.

During that time, a million years had passed since fleshy meat became part of the food of the great ape line. Homo ergaster could sometimes accidentally had their teeth worn away in a pattern that could be caused only by a lot of chawing at flesh. For Homo ergaster, the primary reliance on meat had become key, and in this remarkable species, it was possible to see the great evolutionary benefit of that shift in diet. As source nourishment, meat was about several times more energy-efficient than any vegetarian combination; a herbivorous animal (as well as a mostly herbivorous omnivorous animal) had to eat about a hundred times the weight of plants compared to the weight of meat that would provide the same amount of energy to a carnivore (as well as to a mostly carnivorous omnivore). So the size of the gut needed to process food was far greater for a herbivore than a carnivore.

The complex stomachs and intestines that ruminant animals had were so big that much of the energy they got from their food was used up simply in processing it. For Homo ergaster, its slim skeleton revealed how much smaller its stomach was than that of any other great ape that came before. Instead of a ribcage that widened out as it went down, to encompass a huge mostly plant-eating gut, as in chimpanzees, or the other earlier great apes, the Homo ergaster ribcage remained slender at the bottom, allowing for a small stomach, similar to that of the later modern humans (which would be mostly herbivorous omnivores). A small stomach meant that not much energy was being consumed in processing food, so Homo ergaster's food energy was available to be spent on something else; for there was one other organ in the body that required vast amounts of energy, far more than even the gut: it was the brain.

Running the brain was the most energy-intense process in any animal. For example, brain tissue has 22 times the rate of metabolism of muscles, so one needed

22 times more energy to think than to walkabout. Human brains are nearly five times larger than would be expected of a mammal of their body size, while their guts were only half as big. So it was fortunate for Homo ergaster, and indeed for its descendants, that their ancestors began the shift to being mostly carnivorous omnivores because without that simple act the Hominid brain could never have evolved to the size it has reached during this time.

But within all those statistics there lies the puzzles of what it was that drove the evolution of the larger brain and what made the advanced great ape brain such an advantage to have that it is worth devoting so much energy to it.

The gorging on the mudflat was drawing to a close. The Homo ergaster hunting party had eaten its fill of the soft fleshy meat, and sinews and skin lay torn and sliced around the blue wildebeest carcass. The stone tools were bloodied and worn from the crush against bone. But the Homo ergaster was not yet finished with their kill. Calmly, the older male began to flake the edge of his hand ax with another stone, sharpening the edge, before bringing it down with great force onto a limb joint. Now the blade was brought across the ungulate's hide with a new brutal skill: butchery. Gradually the other Homo ergaster began to do the same, and steadily the remains of the wildebeest were dismembered.

Some way off, a single male lion had begun to approach the kill. The Homo ergaster worked quickly, edgily glancing around them for signs of trouble from the interloper. They knew what a lion could do, but also the Lion had learned that a troop of these great apes was not something to take on lightly. They were scavengers that he could not simply chase away with a show of raw aggression. These primates were clever enough to fight him back, as a team, and so he waited till they had completed their work. And it did not take them long. Within a few minutes, there were hocks of meat lying ready, the older Homo ergaster male had slung a "vest" of meaty skin around his neck, and there were chunks of bone being heaved onto shoulders, as the party began to set off on the long walk back from their kill.

    The big brains of Homo ergaster had provided them with useful new skills and new technology and had changed the way they view their world. But, strangely, none of these would seem to have demanded that the brain evolved to be so large. Scavenging in groups had been perfected in many other animals, and although the new stone ax required substantial intelligence to make, Homo ergaster's large brain was already in existence half a million years before these new tools were invented. Instead, the blue to the large brain lay in what they did with the meat they had gathered. Homo ergaster was one of the first great apes to have developed the skill of disjointing a carcass to take the meat away to share. They had learned that there were advantages to giving what was left in return to particular individuals. Homo ergaster had one of the most complex social relationships of any creature to have walked Earth by that time. As a result, they needed every bit of brainpower that they could bring to bear.

Far from the hunting party, away from the baking mudflat, nestled amid some grassy dunes, lay a small area of wet ground with shrubs and trees. It was an oasis in the dry East African landscape and a place where the Homo ergaster hunting party's group was gathered. These were creatures for which movement was the norm, and here they were gathered to rest, recover from exertions, care for a sick or old member of the group, and awaited the hunting band's return.

It was a busy scene. There were around 20 individuals in this group, but mostly here were the adult females and the youngsters. A Homo ergaster would not be fully mature until it was 12 to 15 years old, and during their long upbringing, the children must learn quickly about the world around them, nor had little chance of survival. In the dunes and scrub around the watering-place, the resident Homo ergaster had been foraging for additional food: anything from the large African insects to bird eggs, fruits (such as berries), and occasional roots. Although fleshy meat was a primary source of energy, Homo ergaster would eat almost anything.

It was the foraging group who heard the returning hunting party first. It was not the sound of their tread nearby that gave them away, but instead, they were heard from much further away. The hunters no longer needed to keep the silence of the chase. Instead, they were free to enjoy the success of their kill and to trumpet the news to others in the group. The sound that the foragers could hear could only be described as singing. The loud calls of alarm and bonding that all primate species display evolved very early in their ancestral line, and "singing" is a common feature in the primate world. But the tall Homo ergaster body had conferred a new ability on to these humans and other primates and certain other mammals. By losing heat through their sweat glands under the skin,

Homo ergaster had freed the lung and airways from the requirement of constant panting in the hot climate. Freed from this constraint, the lungs and diaphragm had evolved to provide a more controlled airflow through the throat and mouth. The result was a new kind of control over the vocal sounds they emit, and in Homo ergaster the rudiments of primitive speech had almost certainly made their first appearance in human evolution.

The faint rhythmic sound was becoming louder in the stillness of the hot evening. Some of the waiting Homo ergaster had stood up and begun to walk towards the echo of the retuning hunting band. Suddenly they were seen, emerging over the rise of a dune. All had returned, and they were laden with meat. An immediate buzz of excitement ran through the group and was now seen as an intricate web of relationships unfolded, revealing the full complexity of the Homo ergaster social life.

There was an atmosphere of friendship between the creatures, as the hunters showed off their booty, and the youngsters flocked to the source of food. The party then quickly broke up as the scavengers paired up with other individuals and moved off to separate parts of the oasis to feed. The old female sought out a younger female in the area who was expecting a child. They greeted each other with grunting hoots and the older one handed over some food; they were clearly mother and daughter, and briefly, they ate together before the older female moved off. The older male strode up to the pregnant female, clearly the father of the unborn child, and gave her some wildebeest meat. She greeted him with grunted affection, and the two sat and groomed each other as they ate, she occasionally scratching caked mud from his limbs, he picking dried blood from her face.

Homo ergaster society marked the beginning of a kind of relationship that had become crucial to human existence: the pairing up of males and females in partnerships.

The evolution of the larger brain, alongside the narrower pelvis that came with upright walking, had meant that infants were now born before their brains had fully developed. Homo ergaster children, therefore, had a very long period of infant dependency while the brain continued to grow, and while they learned the many complex things that were involved in Homo ergaster society. As a result, the energy demanded now placed on a female while rearing her young was huge, so there were tremendous mutual advantages to forming a bond with a partner.

Females knew there was someone to supplement their foraging diet with high-quality meat, and also helped protect their children, who now had such a long period of infancy. In return, males could know for certain whether or not a female's offspring was really theirs, and so it was worth investing their time in support for the child.

This was perhaps a scene unlike any other seen on Earth up to this time. It was a group of animals held together not just by safety in numbers, and not by the domination of a single all-powerful male or female, but by something more collaborative and potentially much more powerful: by the ties of family, and friends. Cooperation had become a powerful force that held a Homo ergaster group together. Underpinning Homo ergaster's social system was a newly emerging idea: trust. They were dependent on being able to understand motives and being able to know that both sides of a bargain would be held too.

Managing this complex world of Homo ergaster relationships was the key to what has driven the evolution of their brain. The size of this full Homo ergaster community was probably a little over a hundred individuals, usually broken up into smaller foraging groups like this one, with the full number coming together in times of plenty, when they would need to interact well together.

This community was a far larger and more complex group than was maintained by any of their ancestors, but along with the bigger brains needed simply to track those other individuals, the consequence of such a large network of relationships had been to kick start another peculiarly future human attribute.

To bond with the other members of the community, the Homo ergaster no longer totally relied on physically grooming each other in the way that the other primates do, there were simply too many individuals for them to have the time to do that. Instead, the beginnings of complex vocal language had emerged, with their rudimentary complex verbal speech enabling Homo ergaster to form friendly and intimate relationships and maintain them across a much wider social community. Thus their upright physique, their larger brains, and their more complex social groups had come together to produce this most significant evolutionary step.

The impetuous young Hoovu swaggered around the group, posturing, standing upright, eager to be noticed by the others of his kind. For him, the kill and his role in it had been particularly important. It meant he had not returned empty-handed and would have some meat with which to woo Doshi, the young female that he wanted to mate with. She, meanwhile, was sitting beneath a tree at the edge of the group, eating berries, with juices running down her chin and chest. There were other younger females nearby and she seemed oblivious at first to the fact that she was the focus of any attention. She smiled as Hoovu walked over to her and hands her some wildebeest meat, and they began to chatter.

These creatures' language of rudimentary words certainly went beyond purely emotional kneejerk calls. They had words and other sounds that were the names of different things as well as each other, words for feelings of happy or sad or fearful, and words that could cover concepts to do with each other. They had a rudimentary word for 'friend'. They were even able to string emotions and thoughts together. But they had no comprehension of syntax or grammar or a tense, just a simple combination of words and other sounds that made sense according to the circumstances.

Doshi reached down and picked up a bird's egg to offer to her prospective mate, exchanging a gift for the antelope meat she had been given and confirming her goodwill with eye contact. It was a sign of the trust that was to build up between them, and Homo ergaster was one of the first great apes to develop a strange new mutation that helped that process of reassurance.

They were one of the first mammals to have visible white sclera or the whites of the eyes. This arose as individuals began to get an insight into the fact that other individuals had thoughts and feelings like they did. This simple feature of the eye had within it considerable power of communication, because it clearly exposed the direction of the gaze, and opened a window into the thoughts of the mind. The intent communicated in the direction of a glance could reveal whether a Homo ergaster was looking into the eyes of a trusted friend or a deceitful enemy.

Suddenly Hoovu started back, looking up into the tree above him and Doshi, and his face distorted into a snarl. There above them, hidden until now by the branches and leaves, was another male, not just an apparent rival, but an interloper to this community. The confrontation was instant and intense, betraying the fact of Homo ergaster's closeness to its primitive origins.

The gentle scene of courtship was transformed into a screaming, gesticulating frenzy of physical activity. No uninvited male was welcome, but for some males (like this interloper) the high-risk strategy of trying to enter another troop and community could bring rich rewards. If they were lucky enough to bond with one of the unattached females, they would be accepted and gained access to the security offered by their new 'family'.

At first, he stood his ground, gesturing, trying to declare his innocence. But the noise and commotion soon bring over others of the group, and he was quickly surrounded by them. His behavior changed, trying to convey penitence, looking away submissively from the circle of angry resident Homo ergaster; but it was to no avail. Suddenly the attack on him came, first, surprisingly, from the old female, and then immediately from the older male, whose great strength was expressed by a massive blow to the interloper's body.

After brief moments of confused movement, the interloper managed to break free and ran off into the bush, while the others shouted after him.

As the group settled to a semblance of calm, with occasional cries of anger echoing across the dunes, Hoovu looked with gratitude towards the older male, who had exercised his natural authority but had nonetheless helped him out of trouble. The group had bonded that bit closer with the successful expulsion of an enemy; the older male masked in the admiration that came his way, but Hoovu had also had his position underlined as a valued member of his community.

The Homo ergaster community group was settled in their area as the sun was setting. But now they had moved on from the baking mudflat and had also moved on time. The cry of a human baby was heard and the older male hunter stroked the head of the newborn infant that he rested on his lap, while his partner was quietly grooming his hair.

Further away the old female was playing with another grandchild; with the help of the grandmother, the female would soon be able to produce more young, although by now she was bonding with a different male. Hoovu and Doshi were now clearly a pair, sitting away from the others expressing occasional simple word-like sounds to each other. Surprisingly, on the very edge of the group, the interloper was now squatting, chanting to himself while knapping at a hand ax.

He was on the periphery of both the group and the community and was not yet fully part of residents' lives, but through persistence and a slow buildup of trust he had become accepted as a presence. He now had security, and the group would have a new strong male to help support their numbers.

In later times future human species would develop amazing in less than a hundred years of their existence.

Homo ergaster and their descendants would come up with nothing new in a million. It was hard to understand such a lack of ingenuity. They simply could not imagine that life could be different, and perhaps it was the lack of creative imagination that in the end marked them out as being not fully advanced.

However, there was one further major advance during that time with Homo ergaster and Homo erectus. By 1.4 million BC, these creatures would encounter fire in a very different way. Unlike all other animals, including all of their ancestors before them, they did not always back away from the sight of a blazing fire. There had been a moment when one of them saw that they could use it. And that single spark of intelligence was enough to transform their lives and to eventually bequeath enormous power to all of their descendants.

They had seen fire before, and came close to it, following natural lightning strikes, or bushfires started by the smoldering of parched grass in the heat of the dry season on a savannah. But this was different. It would possibly be many thousands of generations on before the humans would learn to actually create fire, but it was uncertain which of them it was who shall achieve this remarkable feat. But once the fire was there before them, they shall learn to harness it and to probably control it. Warmth for survival, a weapon for protection against other predatory animals, and even the cooking of food: the advantages of fire were legion.

Fire shall become a basic part of life. As they gathered around the blaze, the humans experienced something of the primeval feeling that later varieties would all still seem to share, as they let their minds wandered, staring into the flames of an open fire.

The Great Migration

Period: Early Quaternary (Early Pleistocene, Early Calabrian)

Date: 1,200,000 BC

Location: Southeastern Asia

Homo ergaster's understanding of the environment within which it had evolved, its ability to plan, to think ahead, to manage the complexity of its social relationships, and the strength of the society that had developed to support it, had all made it the most successful great ape so far to have walked Earth. And it had also been given the confidence to reach out further from its African origins. Homo ergaster could move into another, unfamiliar landscape, and read the signs that would enable it to survive there, and this ability meant that it had come to almost dominating the world, like no other primate species before it.

Although Homo ergaster first appeared around 1.8 million BC, originating in eastern Africa, within less than 200,000 years its descendants had begun to migrate into the Middle East and onto the edge of Europe and to Asia. The new human species that undertook that journeying was known as Homo erectus, coming in 8 different subspecies and was almost identical to Homo ergaster, but with thicker skulls and a more pronounced brow ridge.

Homo erectus had moved across whole new continents and through whole new environments, north up the Nile, up through the Sinai Peninsula, and further north up to the Black Sea. Others spread east, skirting the mountains of the Horn of Africa, across the coastal deserts of the Red Sea, over the land bridge near the Gulf of Aden (this land bridge soon being gone in the future), along the dusty plains of the Palearctic southern coast, across the Indian Subcontinent and around the foothills of the fledgling Himalayas, and into the steamy jungles of Southeast Asia, to make an eventual final stand in Java, an island in Malaysia. Still, others had branched off north, eventually to arrive in the lands north of the Himalayan Mountains.

Some Homo erectus would also eventually make it to the islands of Malaysia and shrunk down into the smallest human species ever by 94,000 BC, Homo floresiensis.

It was a migration over 6,000 miles but, while the journey was indeed epic, in terms of human evolution (given Homo erectus's abilities) it is not in fact as remarkable as it sounds. Subspecies populations grew, so the offspring might need to move on, to ensure plenty of food for everyone. If each generation simply moved its basic camp a mere one kilometer further on, then the journey would easily be accomplished in a shorter time. But the journey was made, and by the time the generations had worked their way across a quarter of the globe, they had adapted to new environments and were thriving across southern Eurasia in 8 diverse subspecies.

Little was different between Homo ergaster and Homo erectus, but for one puzzling thing. The refined stone tools that had been such a feature of the African Homo ergaster lifestyle had never made their appearance with the Homo erectus in Asia (but the Homo erectus tautavelensis, or Tautavel Humans, of Europe, still had them). It seemed that the first humans to migrate took with them the primitive stone tools of the time, and never came to develop anything more sophisticated.

It was the middle of the day, but a shadow hung over the face of a male creature that was treading carefully through the dense undergrowth. Above him were the thick green leaves of a subtropical forest canopy with the light of the sun only fleetingly forcing its way through to the ground.

Around him were the tall stems of a tough flexible wood, which made progress through this jungle slow and painful. He was surrounded by bamboo. This was a world very different from that of the previous Homo ergaster from Africa, but this human was in his familiar element.

At first glance, his was indistinguishable from Homo ergaster, but his species' face was thicker set, measured about 5 feet and 10 inches tall (shorter than Homo ergaster) and in one crucial respect he was different: he had a paler skin, probably reflecting his life in the monsoon climate of southeastern Asia. The ape scrambling through the jungle was a Homo erectus, of the subspecies Homo erectus nankinensis (or the Nanjing human).

Clutched in his hand, and those of the companions (also male) who were moving swiftly behind him, were not flaked stone hand axes, for workable flints were rare in this landscape. Instead, they were carrying the vicious, sharp points of bamboo sticks. Bamboo was a plant that the different southeastern Asian subspecies of Homo erectus had learned to work into a variety of different implements: stabbing points and sharp-edged cutting blades. It was hard and strong and its splintered edges were razor sharp; it well served the purpose of these forest hunters. But unlike stone, it would be immediately discarded once used and would quickly rot away, leaving no trace.

The Nanjing Human in the lead suddenly started back and paused. He drew his hand across his face, wiping away a massive spider web that he had inadvertently walked into. As soon as he had got over the surprise, he immediately looked around for the source. There on the tree was a large fat spider, and in one swift movement, the Nanjing human grabbed it and put it in his mouth. A quick bit and it was gone, with only hairy legs being spat out in distaste. Like Homo ergaster before it, Homo erectus was a culinary opportunist. One of the lead male's companions was stripping a piece of bamboo to create a needlepoint, which he then stuck into the gap in some tree bark and stabbed. He pulled out the bamboo needle, and on the end was a big fat beetle grub. It too was quickly dispatched as a snack.

The voice of the third Nanjing human called, and he rushed forward to join the others, now running, crouching, leaping, across the forest floor, until suddenly they stopped. Crouching down, one of them sniffed at some fresh dung on the ground, left by a small deer. They all remained still, listening, eyes alert, and scouring the forest for signs of bigger, non-arthropod prey. Ahead of them, a female red muntjac (3 feet long and 2 feet tall at the shoulder) stared back, frozen with fear and anticipation.

Bamboo needles, bamboo blades, and bamboo spears: the different Homo erectus subspecies of this region were masters of working bamboo, and the short stabbing sticks in their hands were deadly weapons. Like their African cousins, Homo erectus had become supremely adaptable and responsive to the environments in which they found themselves, and these particular subspecies were as at home in this steamy forest of Asia as were their ancestors in the open savannahs of Africa. Slowly their bamboo weapons were raised for the kill. Once targeted, no prey stood much chance of escape from these humans. Suddenly it was all over. With a crash of splintered wood and broken branches, an enormous creature lunged forward between the Nanjing humans and their prey. It was a male Gigantopithecus, of the species

Gigantopithecus blacki, a huge pongine great ape standing up to 9.8 feet tall. With a great bellowing grunt, he charged the hunting group, who turned, fleeing in panic, allowing one of their numbers to fall and be brushed aside by the charging pongine and end up knocked out.

But the injured human was not left alone, and later on, his companions helped him revive from his unconscious state with water from a bamboo leaf cup. They may not be strong enough to overcome every fellow-creature in the forest, but in the long-term, their intelligence was one of their lifesavers. Indeed, Homo erectus was destined to live on in such places as the Far East, longer than the length of time Homo ergaster had lived. Homo erectus would become one of the longest-living species of humans ever to walk the planet, thriving for over a million years, their last traces dying out around 143,000.

The adaptability and intelligence were shown by Homo ergaster and Homo erectus in moving out of Africa would allow them to possibly dominate the rest of the world for a million years or more. However, there were not too many powers of imagination on their part. Although they had invented and perfected the most advanced stone cutting tool yet seen (the hand ax), they and their descendants would make no new change to its design, and no improvement to its function for all of those ages. For them, the hand ax must have emerged from an evolutionary need to have better tools. But now it worked well for everything they needed, so there was no pressure to do anything different. Their world was settled, so they had no need to change a thing.

In a million years, their technology (and so their brains) would not advance by one degree. It would not even occur to them to do something as futuristically basic as fit a hand ax onto the end of a stick to make a spear.

This would be the next stage of human evolution, the birth of characteristics that really marked out later members of their genus from all other living things on the planet: creativity and imagination.

The Here and Now

Period: Early Quaternary (Middle Pleistocene, Middle Ionian)

Date: 400,000 BC

Location: Southern Great Britain

A male human was seated on a low tree-lined hill overlooking a grassy plain, which was peppered by large boulders. He was staring ahead, thinking. His kind's ancestral relatives, called Homo antecessor, had inhabited these northern European lands for some 350,000 years or more before his kind. His species was the product of thousands of generations of adaptation to the world, in which they lived, and their lineage had survived the bleakest, most hostile, and unpredictable environment that apes had yet had to face. A million years after the long migration of Homo ergaster began out of Africa, archaic humans first reached Europe.

Generation by generation Homo ergaster's descendants had traveled, each community seeking that little bit of extra territory, moving further than the last, hill by hill, valley by valley, and riverbed by riverbed. At one point, their descendants had journeyed slowly and steadily through the Middle East, and to the north around the vast Black Sea. They had traveled northeastwards to the Far East, where they lived on, almost unchanging, until very recent times, like Homo erectus.

But descendants of Homo ergaster also spread into Western Europe, such as the European Homo erectus subspecies called the Tautavel Humans, by a route of which not much was known, and humans as a whole arrived there at about 1.2 million BC. Totally new species began to evolve, each European human species being something very new, such as Homo antecessor and Homo cepranensis.

Still long-limbed and tall, they had become stronger, harder, and smarter: still with the heavy brow ridge and sloping forehead, but now with a larger cranium, and bigger brain that offered far more able to adapt to their new lands. They needed every scrap of cleverness they could muster because, by the time their ancestors were established in Europe, they faced a new pressure: harder weather.

For a reason mainly having to do with Earthmoving a bit further from the sun, beginning at 2.6 million BC the climate of the planet began to enter a period of general cooling, coupled with great instability, alternately cooling and warming with increasing frequency, so that huge swings of climate change occurred, called the Quaternary Glaciation. In Europe, the cold brought ice, with glaciers reaching out for many thousands of years, and then retreating as the climate warmed again, leaving the land scoured and littered with boulders. From 800,000 BC Europe was predominantly a cold world, but there were also times of warmth known as interglacial periods. This was the realm that the European species of humans had come to know, and learned to survive in.

But none of these thoughts were entering the head of our grim-faced hunter. Indeed, he was not capable of thoughts of that depth. His mind was focused on only one thing: his plan for a kill that was about to play out in the trees at the edge of the plain below. Clutching a stone-tipped wooden spear and rising silently to his feet, he signaled to two fellow male companions of his kin, who were also his brothers, and the three men began to move slowly forward, part crouching, through the tall grass.

A short distance ahead, the object of their attention was grazing. It was a male giant deer, and as he continued to feed on grass and bushes in the early morning sun he glanced around, pausing, nervously sniffing for danger.

On their leader's signal, the predators fanned out, being careful to remain upwind of the large deer. Patiently, they stalked him, until finally, they were in range. Hiding behind a huge stone, the leader stood up, arched back his muscular arm, and launched his throwing spear. The others followed suit, hurling a volley of weapons with deadly force.

These brothers were Homo heidelbergensis, and they were perhaps the most recognizably advanced human species yet to have set foot on Earth. They were armed with all the attributes that had proved so successful in hominins up to now: the ability to walk upright, the ability to take a flexible approach to life, and the capacity to thrive within large, intricate societies. They had begun to 'dominate' their world to an extent never seen before. This attack on the giant deer was no opportunistic stalking and scavenging of a dying animal, but a thought-through strategy for killing. These humans were strong, confident in their ability, and unafraid of taking on the largest and most fearsome of other animals as prey.

At that time, the climate of Europe was in one of the interglacial periods that had become characteristic of the planet. The ice sheets had retreated far to the north, and southern Great Britain was a temperate land dominated mainly by large herbivorous animals, or herbivorous megafauna, that grazed on a rich flora, and which were themselves the prey of animals that were strong carnivores and/or omnivores.

As well as the giant deer, there were wild horses, Merck's rhinos, Straight-tusked elephants, and European hippopotamus; there were gray wolves, common European adders, Panthera fossilis, and other felids (Eurasian lynxes and wildcats); there were mountain hares, shrews, voles, bats and common hedgehogs; for birds, there were geese, ducks, gulls, great auks and Old World flycatchers (such as European robins); in the rivers and along the seashores there were Atlantic salmon, newts, common toads, European eels and even Atlantic bluefin tuna. For now, it was a good land in which to live. But it was also full of danger.

    Baying with the pain of the spears sticking into his side, the giant deer broke into a run, but the Homo heidelbergensis remained calm, waiting for the wounded even-toed ungulate to slow. Sure enough, gravely injured, he stiffened and gradually collapsed to the ground. Seeing this, the hunters rushed forward, their faces were a picture of aggression. Their bodies were extraordinarily powerful. With males averaging about 5 feet and 9 inches tall and females averaging 5 feet and 2 inches tall, thick-boned and muscular, they were massively built and fearsomely strong.

As they charged, they called out to each other strange words that meant "Attack! Attack!" They had developed a very basic language of complex vocal sounds, with simple words like nouns and verbs. Its only use was to help them with concepts that were rooted firmly in the present, rather than think about any deeper history. But communicating ideas in the present was an advantage nonetheless. These great apes could think of tactics and use them as a team.

Two of the brothers distracted the great deer at the front while the third, the youngest, who shall be called Dauk, crept closer from behind clutching a heavier stabbing spear and tried to strike the thrashing prey in the neck. But it was a very dangerous move. With one last effort, the giant deer staggered back to his feet, a huge terrified animal almost twice the height of the humans, standing about 6.9 feet tall at the shoulders, with antlers that splayed out a maximum of 12 feet from tip to tip. His weakness meant that his only defense was to wave his antlers from side to side, but that alone was enough to do damage, and the jagged blades audibly swooshed as they curved through the air. Dauk had to jump back more than once to avoid getting hit, and suddenly he lost his footing. He was athletic and agile and was off-balance only for a few moments, but it was enough. With a sickening crack, the large antlers dealt him a savage blow across his head.

Roused to rage by this turn of events, the remaining two Homo heidelbergensis moved in suddenly to finish the herbivore off. The leader stabbed him repeatedly, angry beyond control until the laurasiathere lay dead. Then, chests heaving with the exertions of the chase, the two male humans paused. One bent to his fallen brother-comrade, to inspect the damage done by the deer, while the other set about the skillful task of butchering the dead giant deer with his sharp stone hand ax.

By nightfall, the hunters had returned to their campsite at the edge of the woodland. Here, hunks of meat lay wrapped in the skin from other mammals (as well as collected edible foliage) on the ground, and the group of Homo heidelbergensis was gathered around a burning campfire. They were eating hungrily, and would be relaxed but for one thing: the injured young brother Dauk was semiconscious, blood oozing from the deep gash in his head. He too was wrapped in the skinned hide of another mammal, half sitting, half lying, resting in the arms of a female who was making gentle rhythmic sounds over him, a noise that might one day be called singing. The leader was squatting nearby, chewing herbs, which he packed into the wound. They were clearly anxious and restless because Dauk was a strong young hunter, and the group knew it would be weaker without him.

As the firelight gently died into the night, so did Dauk's life as it slipped away, and the two other Homo heidelbergensis who were tending to him displayed an almost advanced show of grief for the loss.

In many ways, Homo heidelbergensis seemed very like the yet-to-evolve modern humans, with their similar, if large, physique, their skills with tools, their ability to communicate in a complex vocalization, however simply, and even their deep emotions. But there was something deeper that was very different, which is revealed by behavior that to modern humans would seem almost unacceptable, and the following morning it became clear.

Now, around the campfire, there were only the signs of departure. The embers were almost cold and a scrap of wood and flaked stone laid everywhere. Piled to one side were the hacked remnants of the meat, plants, and bones the humans were sharing, but sitting at the edge of the empty circle that marked this particular Homo heidelbergensis occupation there remained only one individual.

Hunched up and still, his head resting on his knees, he was dead. The injured Dauk had been left precisely where he died, his arms and feet partly buried in dirt as a semi-grave, and the others of his group had moved on, carrying their meat, stones, and spears with them across the wooded plain.

For the future modern humans, it would be unthinkable to leave their own dead in this way and to show no reverence for the body of someone who was once a valued biological relative or friend. But for Homo heidelbergensis, it would be unthinkable not to leave him.

The Homo heidelbergensis mind could not look beyond the world of the every day, or beyond the knowledge and memories needed to survive and thrive in the tough world around them. Concepts of who they have existed, but concepts of what might happen to them in the far future, simply did not enter their heads.

Although they cared for Dauk, now he was gone, and there was simply no benefit for them to linger with him. Their world was the here and now, the recent past and the immediate future. Imagination was something that had yet to properly evolve in humans.

Into the Freezer

Period: Early Quaternary (Middle Pleistocene, Late Ionian)

Date: 150,000 BC

Location: Western Europe

As is so often the case in the story of human origins, the climate had been a critical factor in setting the framework for late human imagination to emerge. Around that time, humans had been present on every continent except the Americas and the isolated realms of Australia and Antarctica. The period of interglacial warmth had allowed Homo heidelbergensis to establish itself throughout the northern and western lands of Europe. But the climate had begun to change, plummeting to a new and brutal age of ice cover.

In the northern part of the northern hemisphere (as well similar things going on in the southern part of the southern hemisphere), the gradual arrival of the ice brought back the brutal, harsh world of cold. The warm-weather fauna of the northern portions of the northern continents disappeared, especially in Europe and Asia: Palaeoloxodon elephants gave way to their mammoth cousins, and the hippos gave way to rhinos (Elasmotherium, Stephanorhinus, and Coelodonta).

The temperature fell by an average of about 9 degrees Fahrenheit. But this was just the beginning of a wild ride of the climate. Over the next 200,000 years, the ice advanced and retreated in the familiar cycles of cold and warm. Sometimes each of these times could last for many thousands of years, but the climate was capable of changing from one state to the other very rapidly indeed, over as little as a few human lifetimes.

For the Homo heidelbergensis at the start of that time, the change was extreme: in just a few generations, the land where they had hunted prey across open grassy plains and along wide river deltas became a 'sea' of ice and snow. As the ice advanced and retreated, populations of Homo heidelbergensis abandoned and recolonized the land (some, though, remained in the original human homeland of Africa and became Homo rhodesiensis).

But ultimately, the pressure on these humans meant that they had to adapt and, over 150,000 years or more, their descendants evolved into a new kind of human: Homo neanderthalensis, or Neanderthals.

Around the same time these new humans came about, climate change had brought something to another key part of the human world: Africa.

As the ice swept out across the northern hemisphere, more and more water was drawn out of the atmosphere and oceans to become locked up in the vast ice sheets and glaciers. Water the might once have rained from the skies remained firmly on the land as a solid. So ironically, while the world was getting much colder, in the equatorial regions of Africa, where the ice never reached, it was getting very much drier.

Africa became parched as it could ever become and the climate pressure on the humans, who lived there, the Homo rhodesiensis, was very different, but equally as great. And from this drought-ridden land, yet another kind of human emerged: Homo sapiens, or modern humans.

Of these few great species, only one would ultimately survive. By that time it was by no meant certain what the outcome would be, for the Neanderthals were very successful indeed.

Europe during the Last Glacial Period (or ice age) was a bleak land. The local ice cap, known as the Barents-Kara Ice Sheet, almost permanently covered most of the region, reaching south as far as southern England, creating a blanket over a mile thick. The glaciers of the Alps spread out like tongues of cold across the central lands of France and Germany.

In between the sheets of ice lay vast swathes of the frozen tundra that stretched from Denmark all the way to the French Riviera. It was a forbidding and unrelenting landscape where only the hardiest forms of life survived, and the work required to stay alive was formidable.

Three male Homo neanderthalensis, or Neanderthals, were picking their way along a vast lagoon of blue ice. They were wrapped in the furs of fellow mammals and were carrying wooden spears.

15,000 more generations under the evolutionary pressure of extreme climate change had transformed these descendants of Homo heidelbergensis. No longer as tall and as long-limbed like their now-extinct Homo heidelbergensis relatives, Neanderthals had evolved a shorter, stockier, altogether more solid physique that was well-suited to their freezing environment.

Males stood 5 feet and 5 inches to 5.5 feet tall and females about 5 feet to 5 feet and 1 inch tall. This shorter, rounded shape provided less surface area for the body to lose valuable heat when the temperatures around them could regularly reach as low as 30 degrees Celsius below freezing.

Here, the summer was drifting into autumn and the region was already blanketed with snow. These Neanderthal men left their group's camp the day before in search of food, and nothing had materialized until now. The leader of the three, who shall be called Zoy stopped, turning his head that way and that, listening, and anxious, occasionally giving a rasping cough. With winter conditions setting in so early this year, many of the other animals that they would normally hunt had already moved south, and the meat was now scarce. Neanderthals did eat plants as well, but they were mostly carnivorous omnivores. For Zoy, he was in a dilemma: he knew that they should move south with the large prey, but he also wanted to stay in the hope of finding food, because for him the search had particular importance.

Back at the camp was a young Neanderthal woman, heavily pregnant with his child, and he was worried that the rigors of the journey could kill her or the baby if they tried to move on. But he also knew that he must find her some food soon, or risk losing both her and the child anyway.

His companions were unhappy with his determination to stay, and they told him so. Neanderthal thinking and language (mostly with their vocals and partly with physical gestures) had now progressed enough to express far more complex ideas than the Homo heidelbergensis who lived here so long before:

adjectives, question words, qualifiers, and more were all in their repertoire.

One of his companions said that there were certainly no signs of prey around, but it would be dangerous to remain here, so the offer was giving him company. Ideas of the future and consequences of actions could all now be voiced, and argued, for the Neanderthal brain was large, with an average cranial capacity of 1600 cubic centimeters. It was even larger than that of a modern human, although perhaps not in the crucial area of the frontal lobes, where a Modern Human's most complex reasoning was carried out. These humans were not stupid creatures; they could see a problem that mattered to them and disagreed about how to deal with it.

A decision to go hunt now in the lowland valleys to find reindeer (also known as caribou) together was being made. Zoy was determined to stick with his decision to stay in the current area.

Coughing again, he headed off down the slope and came upon a snow-covered bush. He was immediately cheered by the idea of berries and called out. But the smile vanished from his face when he shook off the snow to reveal only the shriveled remnants of old fruit.

Then suddenly, before the others had time to protest, there was a flutter of movement in the snow, and the chase was on. The three heavy-set hunters had glimpsed prey, and their minds switched instantly to the task they knew so well.

The hunt was all-consuming, and their instant adoption of strategies to catch their victim so instinctive that they barely noticed that the object of their attention was only a mountain hare. However small, it was still food.

The Neanderthal men were now crashing through a sparse wood of shrunken birch trees, with the hare still out in front. As they ran through the icy air, clouds of vapor billowed from their noses, which were the dominant feature of their faces.

Wide and protruding, their large noses were one part of their bodies that were very efficient at losing heat. The result was that during sudden exertions like this, their bodies never overheated, and so they sweated much less. That was another finely tuned adaptation to their world because in a cold climate such as this their sweat would quickly freeze to their skin.

Briefly, the three Neanderthals came to a halt, catching their breath, looking around them, listening for any sign of the hare. Then they were on their way again, moving rapidly down another slope after the hare.

Half slipping, half leaping, Zoy tripped and tumbled through the undergrowth to the bottom of the slope, landing with a shout from physical pain. He scrambled to his feet and looked at his left hand, his pointing finger of which was bent at a gruesome angle, clearly dislocated.

Without a moment's hesitation, he simply snapped it back into position, and then looked up, panting and coughing. Momentarily he smiled. His companions had caught the hare.

Later on, as the three sat around a small flame while the mountain hare gently cooked, their faces clearly showed the scars of many years of physical exertions, and their limbs revealed the bumps and twists of old broken bones. Indeed, Neanderthal life was physically very demanding, and mental strength was just as important.

The hare had made a great meal, but the day was drawing to a close, and the three male Neanderthals were huddling down in their furs, pressed against a rocky outcrop, the flames keeping them secure for the night. In the morning they would press on. They were still concerned for the others who awaited them at their main camp.

The home camp of Zoy and his two companions was a classic overhanging cave in a limestone cliff, with a fire flickering in the mouth. Wrapped in other mammals' furs, sitting alone in the dark, was the pregnant younger partner of Zoy. She was using her time by securing a stone blade to a long shaft of wood. Neanderthals had now perfected the hafting of sharp stone points to their wooden spears.

Inside, an old Neanderthal woman was checking the hair of a younger female for lice. The close physical contact of grooming is a characteristic of all primates that has remained very strong, passed down from some of the most ancient of species, dating back to the Paleogene epoch of the Early Paleogene period.

The younger female herself was scraping the last bit of dried meat from an old piece of hiding with her teeth. Some of the other members in the group were preparing for sleep, by spreading out other mammals' furs on the floor of the cave. The ground was filthy, littered with the detritus of Neanderthal life.

The size of this group was small, with only eight individuals in all, including the three hunters out in the field. This may seem like a small number, which would expose the group to danger if it lost any of its members, but in fact, it was a pattern that had been found repeated wherever Neanderthal activity had been performed. So it was clearly an adaptive response to the world in which they found themselves. Such small numbers in social groups meant that the ties that bound the individuals were strong.

They all depended on each other for survival, so the chance of major disputes between competing individuals was small. But it also meant that during very hard times, such as in the depth of this ice age, what little food they could find did not have to be shared between too many mouths.

This evolutionary strategy for survival in hard times, however, would one day be a factor in this species' final extinction.

Meanwhile, out in the snow, as the moon sank below the horizon and the sun rose again on their temporary shelter, the three Neanderthal hunters were about to catch a much more spectacular creature. Their day was again spent tracking and scouting, as Zoy grew desperate to find substantial food in this very bleak landscape. He could foresee the consequences of failure and was becoming increasingly anxious at the thought of his leadership being challenged, and his cough had worsened.

By nightfall, nothing beyond the carcass of another Mountain Hare, itself the victim of the extreme cold had turned up, and the three Neanderthals spent another uneasy night beneath the stars. Zoy knew that it would take more than another lagomorph to counter the arguments of his companions.

The following day, the ritual was repeated, and by mid-afternoon they were standing on high ground, overlooking a wide valley, scanning the seemingly empty landscape for any sign of suitable prey. But the younger hunters had decided it was pointless. They felt as though they must quit the fruitless search, return to the camp, and move the party to the more plentiful territory.

In the fading light of the afternoon, they turned and began to walk away. Reluctantly, Zoy had to agree, lingering to take one last look across the misty landscape before he began to follow them. Suddenly, from far across the valley, he registered a faint trumpeting sound: a distinctive call he had been longing to hear.

A small herd of six Woolly Mammoths slowly came into view out of the evening mist. The Neanderthals were all suddenly attentive. There was no time for argument or rebuke. They now had a job to do: they had to form a careful plan of action, for these were herbivores that could not simply be tackled at random.

Up to 4.4-6.6 short tons in weight, and with very long and curved tusks 5.9-14 feet long, a Woolly Mammoth could skewer or crush an unwary Neanderthal in a moment. It was an adversary that the group might hesitate to take on, but times were hard and this was an opportunity to get almost as much food as they could wish for.

What is more, all the signs were that an attack could be successful, for the three Neanderthal hunters realized at once that the mammoths were moving towards the end of the valley where it narrowed into a small, steep-sided ravine, the perfect spot for an ambush.

The Woolly Mammoth herd, led by an old matriarch, was oblivious to the three Neanderthals looking down on them as they drew near. The predators did not need to speak, as the plan formed clearly in all their minds. They ran to the edge of the ravine and, as they reached it, the two younger Neanderthals hurriedly looked for rocks big enough to roll down onto the proboscideans below. With luck, one of them would injure a mammoth sufficiently for them to move close for a kill.

Zoy lay down amid an array of small rocks, ready to throw them at the herd when the time was right. He acted as a lookout, while the other two found large boulders near the edge of the ravine and laid on their backs with their feet straining up against the rocks, ready to push them over at their leader's signal. His face was peering over the edge; the tension was palpable.

With surprise on their side, they stood a good chance of success, but they would only be able to strike once. It was crucial that he chose the right moment.

For the mammoths, this one of many familiar routes that the matriarch chose as they trekked south in search of warmer climates, but she was becoming edgy as they entered the narrow confines of the ravine.

Above them, the three Neanderthals hardly dared to breathe for fear of letting this opportunity slip through their hands, but as the large herbivores were almost there, Zoy let slip a small cough. His companions looked at each other in alarm.

For a brief moment, time seemed to stop. At first, the mammoths did not seem to have heard the telltale noise, but it was only a few seconds before they reacted (30 tons of chaos broke out). The matriarch repeatedly bellowed in alarm while the rest of the herd members hemmed in by the narrow rock walls and confused as to what has happened, began to turn around and collide with each other in the ensuing panic. Zoy held his nerve.

Quickly getting back up on his feet, he glanced from the boulders to the herd below, trying to judge again the moment to strike.

By now the Woolly Mammoth matriarch was beginning to lead the herd back out of the ravine, and Zoy was faced with an erratic moving target, but he knew that he had no choice but to attack now or lose the opportunity forever. He turned and gestured sharply to his companions, and with a concerted heave, each launched his boulder over the edge of the ravine and into space.

Again there was a moment that seemed to last forever as they waited for the effect of the attack to become clear; and then celebration: one of the large boulders had struck a mature female mammoth on the hip. She wailed in pain and collapsed onto her haunches. Immediately the humans began to pelt her with smaller rocks until she became dazed and panicked. She irrevocably damaged her broken hip by thrashing around in vain attempted to get to her feet and flee. The Neanderthals stopped the onslaught. The mammoth was not dead, but they knew that they could afford to wait. Time would be their weapon now.

As the deepening gloom of evening beard down on them, Zoy rapped out an instruction to his fellow hunters: they used fire. The rest of the Woolly

Mammoth herd had fled from the valley and it was time for the Neanderthals to finish their work. One of Zoy's companions bent to the ground and, with flint and dry wood, soon had a small blaze aglow. The three Neanderthal men clambered down the ravine, with Zoy brandishing a flaming torch as they approached the exhausted mammoth.

While he distracted the terrified afrothere with the threat of fire, the two others carefully avoided the flailing tusks and stabbed her repeatedly in the neck, and flanked with their long stabbing spears. Mercifully, the kill was over quickly as one of the spears' thrusts succeeded in piercing the prey's main artery. The Neanderthals were jubilant.

The three Neanderthal hunters returned to their home camp, and sat around the fire with the rest of the group in the overhanging cave, enjoying the spoils. Zoy, still coughing, was cutting off pieces of meat with a sharpened flake of stone, while his younger female partner sat by his side grooming him. The two young male hunters had now relaxed all criticism of their leader forgotten.

One of them was humming quietly to himself, as he chewed, while the other ate greedily, with a piece of mammoth meat in both hands.

For Zoy, the successful hunt meant the end of his anxiety in the short term. The meat that the group did not eat now could be kept for a long time in the freezing snow, so they had plenty of food to last them until long after the birth of Zoy's child. When that was over, the group would be able to move on to better land. Zoy himself was happy.

Suddenly the greedy young hunter gulped and choked. The other Neanderthals turned to look at him in alarm, but after several moments of panic, with his eyes staring, and his face bulging and red, the offending lump of meat spewed out of his windpipe and flew across the camp, striking another member of the group in the eye. The whole group burst into fits of laughter, which echoed out of the overhanging cave and down into the valley below. Neanderthals were creatures who had time in their lives for more than simply the process of survival.

Despite the obvious differences between modern humans and Neanderthals, there was a lot about Neanderthal behavior that was similar to modern humans and other great apes: their joy at being reunited or their contentment at feeling warm and well-fed. They had an awareness of themselves as living beings and their place in their world; they understood what was normal, what was routine, and what was unexpected. They laughed at themselves and enjoyed a social life for its own sake. For many thousands of generations, they had done little with this intellectual power, for their world had always been rooted in the intermediate reality of their daily lives. But their potential for creativity and imagination was far greater than perhaps they themselves even realized.

Although Neanderthals were being pushed to the limit by the current ice age, they were supremely well adapted to their world and were destined to survive the harshness of the European/Eurasian climate for a further 122,000 years. Meanwhile, at the same time as they enjoyed their heyday in the north, in Africa, their close cousins were experiencing climatic hardship of a very different kind. But in so doing they set the stage for possible eventual success by Homo sapiens.

Scorching Sands

Period: Early Quaternary (Middle Pleistocene, Late Ionian)

Date: 150,000 BC

Location: Eastern Africa

A female human, covered in a fine layer of dust, was digging a hole in the sandy soil with a digging stick. She worked down a few feet, and then paused and moved on, to start another.

Behind her were dozens of these holes. She was digging patiently for water along the dried-up bed of a river that once flowed through this valley (it would flow again when the rains arrived). She was also hoping to turn up a tuber because these and the remaining pockets of moisture in the river were the only sources of water she could rely on during the dry season. She was not optimistic, but she had to persevere, as she desperately needed water to allow her to produce milk for her infant child.

As she wended her way along the length of the dried riverbed, she passed the skeletons and rotting remains of a few nonhuman animals that died on its banks: gazelles, African bush elephants, and Nile crocodiles.

This was a drought in its most extreme form, and it had been part of this human's life since as far back as she could remember. The icecaps at the poles had grown so large that much of Earth's water was now trapped in ice sheets and glaciers. Less moisture was being drawn into the atmosphere, and so, globally, there was much less rain. And this was most noticeably felt in sub-Saharan Africa, where the populations of such creatures as humans were experiencing hardship like none that their ancestors had ever faced.

Along the side of the creek, just shaded from the direct heat, the female human's group sat huddled together out of the sun. Flies played about their faces with impunity; the entire group was too weak to bother brushing them away. The woman gave up her fruitless digging for water and returned to the others. As she approached, the others looked up at her with pity. An older female got to her feet and handed her the body of her lifeless child.

These humans were the African descendants of the descendants of Homo ergaster. While Homo heidelbergensis had flourished in the north (in Europe), something very like it evolved in sub-Saharan Africa, known as Homo rhodesiensis. But in turn, Homo rhodesiensis (which were African descended cousins of Homo heidelbergensis) had given way to a new human species: Homo sapiens, or modern humans, this early subspecies called Homo sapiens idaltu, first appeared around 200,000 BC. Physically, these early Homo sapiens were almost indistinguishable from the later subspecies Homo sapiens (or anatomically modern humans). Virtually gone was the heavy brow ridge; the face was flatter, the small nose was there, and the chin had clearly formed, and above all a high forehead sweeping up to form a cranium that housed a large brain, having a brain capacity of 88 cubic inches. These creatures' survival was ultimately tested by the extreme dryness of the climate, and the test drove them almost to extinction.

The population Homo sapiens idaltu here in Africa had been reduced to perhaps a few tens of thousands. The reason for this, remarkably, came from their genes. It seemed that this species faced what was called a 'genetic bottleneck,' out of which they shall live. There was no certainty as to the cause of this evolutionary pressure point, but it might have been linked to the extreme climate change around the glacial maximum of this time.

Across the continent of Africa, small populations of these early modern humans struggled through the arid conditions. Some would not make it, but others would. On the margins, some had been fortunate enough to live in local areas that provided liquid water, vegetation, and prey, and these humans shall carry on unaffected; but for many others, the numbers in their small groups were dwindling dangerously low. Yet the effect of that extreme climate pressured on these modern humans may create the very attributes that shall one day make their species so successful. This was like the process of evolution turned up to maximum speed, because when a small population was under pressure only the very fittest survive, and in drought-ridden Middle Pleistocene Africa, fitness meant living by their wits. The difference between life and death could come down simply to a good idea.

Far on from the huddled group of dying Homo sapiens idaltu, over a mountain range, across valleys, past other dried-up rivers, and on through many generations of time, another tiny group of Homo sapiens idaltu was walking, foraging as they went, but clearly on a journey. They were a mated couple and their young daughter, and they were patently exhausted from a morning of toil. The mother then approached a very dry-looking tree, to dig around one side of the base, until she unearthed a round white object and drew it out..

It was a very large ostrich egg, but it was not going to be food for her and her family. She removed a plug of grass stuffed into a small hole, raised it to her lips, and drank water. This was something that she and her partner placed there weeks ago, perhaps even months, when there was some rain. They knew that one day they would be passing this way again, and when they did, this water might be just what they needed.

A simple but lifesaving idea such as this had

marked the survivors out from those whose family bloodlines had come to an end, but it also showed what was special about the modern humans that shall survive. This was not simply the spark of lateral thinking needed to use an ostrich egg for a very different purpose. This was forward planning. They could think of what might happen in the future, and work out a way of dealing with it.

When the climate eventually changed for the better, the modern humans that shall come through the bottleneck were the ones who were fittest not just physically, but mentally. They were the ones who could think their way out of the crushing circumstances they were in.

They were the ones who had developed what one might call an imagination. And the small population size meant that this mental attribute very quickly, in evolutionary terms, had become a feature of their populations.

As the couple and child continued their journey, the woman began to lag behind. The man turned and chided her, clearly wanting her to catch up. She in turn rebuked him with a short stream of angry words that their distant future generations would not recognize. But words they were, for these survivors of the climatic squeeze on the planet, have developed the ability to share their imagination, their plans, and their points of view with others.

Around that time, modern humans had begun to master the art of language, with a subtlety far beyond the Neanderthals communication of needs, actions, and immediate concerns. Human imagination had brought with it an ability to think far outside their immediate circumstances, to think in metaphor and analogy, and the use of language enabled humans to use a symbolic expression. It enabled them to work with a wider range of other humans than ever before.

With a shake of the head, the man was silent, and the trio resumed their journey. They had been walking for the best part of a day in search of something that would guarantee their survival: not food or water, but a mate for their daughter.

Arriving at an open sand flat, they found another, a larger party of Homo sapiens idaltu waiting for them, and for a moment the two groups eyed each other warily, weighing up the situation.

Finally, the male leader of the larger group broke the tension, stepping forward in greeting and speaking to the new arrivals. The father stepped forward and gestured for the young girl, his daughter, to come forward and in turn, the leader of the large group called for a young boy to step out.

This is how a simple union between neighboring family groups was arranged. But whatever the ceremonial procedure, the advantages of such a union were clear to both sides. Each would now have allies to rely on, and so would stand more chance of survival when times became hard again, as both lead males knew they would. As the meeting carried on into the evening, the two men conversed and agreed to the terms, promises of help, which would be to their mutual advantage.

Complex verbal language had become how whole modern human social groups could create and share a web of culture; it was a kind of social glue. At the most primitive level, this ability to share information and common experience, to share the future with another group, and to widen the immediate family, meant that the species could exchange more genes, more quickly, and became ever stronger as a result.

So the stage was set for the final epic moments in the human timeline. To the north in Western Europe and in the Middle East, such species as Neanderthals were the end stage of evolutionary adaptation to the harsher colder climate. To the south in the heart of Africa, it was such species as modern humans.

In the next 50,000 years, Homo sapiens sapiens, or anatomically modern humans, emerged in Africa out of Homo sapiens idaltu, armed with their sophisticated intelligence and ability to adapt possibly any environment, and they started to move to other areas of Africa, as well as into Eurasia 57,000 years after that. The effect on the populations of other human species that already existed in particular Eurasian territories shall be perhaps inevitable.

In the Far East, pockets of different Homo erectus subspecies, the longest surviving species of human ever to have lived, shall survive until 143,000 BC, long before modern humans arrived in Eurasia. But it was in Europe that the flowering of modern human success as a species would be most graphically displayed.

Close Encounters

Period: Early Quaternary (Late Pleistocene, Late Tarantian)

Date: 55,000 BC

Location: Middle East

A low summer sun threw up shadows across the rolling trees of the Middle East. A flock of birds was beating steadily westwards, their backs to the rising sun. It was warm midday and the air was thick with undulating swarms of midges. Underfoot the vegetation was lush with a range of sedges, club moss, and dwarf birch scrabbling for room among dense swards of true grass. Here and their clumps of trees such as larch and alder grew in the valleys. The region was covered by coniferous forests and surrounded by huge mountains.

A male human, covered in mammal hides, was treading carefully through the dense undergrowth. Above him were the thick green leaves of a temperate forest canopy with the light of the sun only fleetingly forcing its way through to the ground.

Around him were the tall stems of a tough flexible wood, which made progress through this forest slow and painful. He was surrounded by trees. The respect he was different: he had slightly darker skin, probably blending his life in the temperate climate of the Middle East. The ape was scrambling through the forest.

Clutched in his hand, and those of the companions (also male) who were moving swiftly behind him. Homo sapiens are believed to have migrated through the temperate forests of the Middle East on their way out of Africa at about 60,000 BC.

A Homo sapien's knowledge of the environment within which it had evolved, its ability to plan, to think ahead, to manage the complexity of its social relationships, and the strength of the society that had developed to support it, had all made it the most successful great ape so far to have walked Earth. And it had also been given the confidence to reach out further from its African origins. Homo sapiens had the ability to move into another, unfamiliar landscape, and to read the signs that would enable it to survive there, and this ability meant that it had come to almost dominating the world, like no other primate species before it.

Generation by generation They had traveled, each community seeking that little bit of extra territory, moving further than the last, hill by hill, valley by valley, and mountain by mountain. It was a migration over 6,000 miles but, while the journey was indeed epic, in terms of human evolution (given Homo sapien's abilities) it was not in fact as remarkable as it sounds. Populations grew, so the offspring might need to move on, to ensure plenty of food for everyone. At one point, Homo sapiens had journeyed slowly and steadily through the Middle East, and to the north around the vast Black Sea. They had traveled northeastwards to the Far East, where they lived on, almost unchanging.

If each generation simply moved its basic camp a mere one kilometer further on, then the journey would easily be accomplished in a shorter time. But the journey was made, and by the time the generations had worked their way across a quarter of the globe, they had adapted to new environments and were thriving across southern Eurasia.

The reason why they left Africa had something to do with climatic shifts that were happening around that time. There was a sudden cooling in the Earth's climate driven by the onset of one of the worst parts of the Last Ice Age. This cold snap would have made life difficult for our African ancestors and the genetic evidence points to a sharp reduction in population size around that time. In fact, the human population likely dropped to fewer than 10,000. These migrations eventually led the descendants of a small group of

Africans to occupy even the farthest reaches of the Earth.

The Middle East back then had a huge and amazing diversity of fauna and flora. As well as deer, there were wild horses, hippos, wild cattle, wooly rhinos, and wooly mammoth; there were gray wolves, Panthera fossilis, and other felids (Eurasian lynxes and wildcats); there were Mountain hares, shrews, voles, bats and Common hedgehogs; for birds, there were geese, ducks, gulls, Great auks and Old World flycatchers (such as robins and masked boobies); in the rivers there were, newts and common toads. For now, it was a good land in which to live. But it was also full of danger.

The lead male suddenly started back and paused. The human crouched down and plucked a couple berries from a nearby bush and stuffed them in his mouth.

The green paradise of the mountain had blossoms and buzzing insects. The sun diffused in daggers of light as the lead male climbed over a fallen tree, ducked under some low branches, and stopped to catch his breath. He listened intently. Took another step forward, and then heard a twig snap.

The voice of the third of the humans called, and he rushed forward to join the others, now running, crouching, leaping, across the forest floor, until suddenly they stopped. Crouching down, one of them sniffed at some fresh dung on the ground, left by a red deer. They all remained still, listening, eyes alert, and scouring the forest for signs of bigger prey. Ahead of them, a female red deer (3 feet long and 2 feet tall at the shoulder) stared back, frozen with fear and anticipation.

These Homo sapiens were perhaps the most recognizably advanced human species yet to have set foot on Earth. They were armed with all the attributes that had proved so successful in hominins up to now: the ability to walk upright, the ability to take a flexible approach to life, and the capacity to thrive within large, intricate societies. They had begun to 'dominate' their world to an extent never seen before.

Suddenly it was all over. A loud yell was heard and it echoed through the trees. The red deer was spooked and ran off into the forest. One of the Homo sapiens ducked for cover. With a crash of rustling leaves and twigs, another human species, a Neanderthal lunged over a log, over the humans. The three of them watched as the Neanderthal ran through the dense forest. They had no idea of what just happened, but clearly that these humans never saw another human species quite like the Neanderthal.

Neanderthals had been occupying the Middle East long before Homo sapiens arrived. They ended up co-existing with the Homo sapiens. Living in small groups, the Neanderthal population had become a collection of isolated pockets. And the Neanderthals thought of their new neighbors in some ways quite like them but in some ways very different. But with their far-reaching networks and strange customs, and with so many more of them arriving with each new generation, they could seem quite terrifying.

The home camp of three humans was a classic overhanging cave in a limestone cliff, with a fire flickering in the mouth at the edge of the woodland. The three hunters returned to their home camp around dusk and sat around the fire with the rest of the group in the cave. One of the hunters was examining a pearl that was dropped by a Neanderthal that morning. It is believed that Neanderthals wore jewelry like their cousins.

But there was a sudden noise that alerted his troop and he quickly lifted himself up, to see the commotion. Just outside their cave was a little girl. She was not any little girl. She belonged to the species of Neanderthal. She was lost and had nowhere else to go. She lost her family. The other Homo sapiens raised their spears toward the intruder. The little girl pleaded. Finally, the male leader of the larger group broke the tension, the adult female Homo sapien stepped forward in greeting and speaking to the new arrival, comforting her. The little girl was accepted by the family.

Ten years passed, the Neanderthal girl, now old enough was cradling a baby. She had recently given birth to this 3-year-old baby. The rest of the Homo sapiens returned from a hunting trip, happy to see the Neanderthal girl. One of the Homo sapiens came to the Neanderthal girl and he glanced over the baby cradled in her arms.

Many Neanderthals interbred with Homo sapiens. Today, two percent of our genes come from Neanderthals. Expansion of modern humans out of Africa can be proved by the interbreeding between modern humans and Neanderthal.

Even though humans interbred with Neanderthals, a major portion of the genome is from

Africa. As Homo sapiens presence expanded all across Eurasia, they interbred with other neighboring species of Homo (like Neanderthals or Denisovan, who became extinct a few thousand years later.)

New Hunters

Period: Early Quaternary (Late Pleistocene, Late Tarantian)

Date: 43,000 BC

Location: Northwestern Europe

Shapes appeared between the high grasses, which gently swayed in the wind. Four Homo sapiens, or anatomically modern humans (being of a particular race with tannish skin), covered in nonhuman mammal skins and brandishing spears were on a hunt. The leader of the four in particular, who shall be called Dirket, had his eyes on something: prey. He was leading the three other members of his party.

For all members of their genus, their unique ability to use stone tools, developed from their African ancestors, allowed them to make the most of what was available. Despite the lack of fur, they wore clothes made of animal hides to keep them warm in the ice age climate.

Dirket signaled the other hunters behind him. They followed Dirket's lead, crawling up the incline, hidden in the grass. Homo sapiens were skilled hunters that specialized in hunting larger prey. Over the rim was a stunning herd of caribou, grazing on an open plain. More than a hundred dominated the landscape. The hunters shared anxious looks. None of them had ever seen a herd this big.

After sizing up the situation, Dirket turned to the others and gave a hand signal. All of the other hunters followed his lead, as he moved towards the herd. The younger male, Roluuk, immediately started to move faster than the others, getting a bit ahead of the group. Dirket made a clicking insect sound, getting Roluuk's attention. Dirket motioned for Roluuk to ease back, to stay tighter with the advancing group. Roluuk reluctantly slowed, waiting for the others, joining with them, and then pacing himself to stay with them. The hunters advanced on the herd. Dirket led.

Even though caribou were near-sighted, their senses of smell would help them locate predators. Their greatest asset was their large antlers that could be used for defense. They picked up their scent and began to stir. Dirket signaled the hunters. They stopped in a long, curving row. Dirket continued, alone, moving deeper and deeper into the herd. He passed the female caribou and her calf, very close. The calf watched him curiously. Dirket zeroed in on the caribou the lead bull. Dirket carefully approached.

Sensing movement, the lead bull turned menacingly towards the approaching Dirket. The other caribou in the herd reacted to the lead bull's movement, and they turned as well. A pause, then they began grazing again. Dirket crept closer still. The lead bull stopped eating. He sniffed the air and the animal made a rumbling sound. The lead bull lo turned towards Dirket, trying to spot him with near-sighted eyes.

Roluuk eased forward a bit. Some of the other hunters noted Roluuk movement. They were not pleased. Suddenly, directly in front of the lead bull, Dirket leaped to his feet, waving his arms, shouting wildly. The lead bull's feet tear up the ground. His massive head whipped back and forth. Dirket stood his ground. He went closer to the lead bull and thrust his spear at the animal's face. The caribou charged. Dirket rolled out of the way, barely avoiding being trampled. Dirket blew his hunting whistle made out of mammoth ivory. At that signal, Roluuk and the other hunters leaped to their feet, waving their arms and shouting. The rest of the herd, spooked and furious, snorted some rearing up on their hind legs.

This attack on the reindeer herd was no opportunistic stalking and scavenging of a dying animal, but a thought-through strategy for killing. These humans were confident in their ability and unafraid of taking on the largest and most fearsome of other animals as prey. They had planned their move carefully refining every detail.

The herd charged after the lead bull with a

sound like rolling thunder, scraping off the earth as they ran. Their stampede led them to the far side of the plains.

Far in the distance were two tall rocks that created a slightly narrow opening. Dirket joined the other hunters, running, giving them hand signals, deploying them.

The hunters formed themselves into a crescent, behind and to the sides of the running caribou, herding them, directing the big animals towards the rocks. Roluuk ran alongside the herd, point man of one side of the crescent. Roluuk, focused only on the caribou, and not on the other hunters, ran faster, putting distance between him and the others.

Dirket saw Roluuk and called out angrily to him, motioning him back with the group, but the thunderous sounds of the caribou's stampede drowned out Dirket's voice. One of the hunters looked angrily ahead at Roluuk. The others saw Roluuk ahead of the group and exchanged an angry look.

Picking up speed, Roluuk got further ahead of the other hunters. Roluuk was now running ahead of some of the trailing caribou, and some of those animals veered to the outside of Roluuk, driving him in among the main herd. Dirket saw this happen. His anger grew, but there was nothing he could do about it.

With Roluuk running among them, two of the caribou in the center of the herd were spooked and started to veer off, away from the mouth of the hills. Other caribou in the herd responded and started to veer away in the same direction as well. Dirket saw the herd starting to change direction because of Roluuk. Dirket motioned to other hunters who tighten up on their side of the herd, yelling, driving the herd back toward the rocks ahead. The lead bull had passed the boulders, and now every animal had to choose its route.

Roluuk ran between two animals who wanted to squeeze between two of the giant stone boulders. They scraped the rock and their bodies crashed together.

Roluuk's only chance was to duck down and dive under the body of one of the stampeding reindeer. For a couple of endless seconds, he was under the animal. But when the caribou jumped over a smaller rock, which was too high for Roluuk, he had no choice other than to jump aside, tumbling to the ground, where he had nearly trampled by the hooves of the other caribou. Two hunters shot him harsh looks as they passed. Roluuk scrambled up, but he had lost his lead. Dirket looked over, angry, but relieved that Roluuk was alive. They ran on, now with Roluuk behind the other hunters.

The other hunters continued to herd the caribou across the plain. Just beyond the other side of the plains were other hunters in the same party. They were already crouched down and waiting for the herd to come by. They saw the herd coming. Derkit blew his mammoth whistle and the other whistled in response. One of them signaled the other hunters behind him. They got up and readied their spears. They used their atlatl, or spear thrower, to toss their flint-tipped spears.

They were the revolutionary at the time. Spear thrower designs may include improvements such as thong loops to fit the fingers, the use of flexible shafts, stone balance weights, and thinner, highly flexible darts. Darts resembled large arrows or small spears and were typically from 1.2 to 2.7 meters (4 to 9 feet) in length and 9 to 16 millimeters (3/8/" to 5/8") in diameter.

It may consist of a shaft with a cup or a spur at the end that was supported and propelled the butt of the dart or spear. The spear-thrower was held in one hand, gripped near the end farthest from the cup. The dart was thrown by the action of the upper arm and wrist. Acting as a lever, it launched spears twice as far.

The apes threw their spears ruthlessly, creating a hailstorm of spears. The caribou herd changed the other direction away from the other hunters ahead. The other caribou tried to keep pace, but they could not go any further. They were soon left behind and were separated from their main herd. One of the spears hit one of the caribou in the hide and they came crashing on their side into the dirt. The other ran exhaustingly toward their herd.

The hunt was a success for these human hunters. They would now have to take what they need as soon as possible, before scavengers like gray wolves, Eurasian cave lions, and spotted hyenas, wanting to scavenge their kill, made this a dangerous place to be for the humans.

Working in groups, armed with formidable weapons and using sophisticated hunting techniques/tactics, these highly efficient predators would quickly spread across the grasslands and indeed the whole continent.

The Future

Period: Early Quaternary (Late Pleistocene, Late Tarantian)

Date: 40,000 BC

Location: Southwestern Europe

Nighttime in southwestern Europe. A male Homo sapiens sapiens, or anatomically modern human (being of a particular race with paler skin), was busy making a lamp, twisting a piece of roe deer sinew around using his teeth to make a wick. He poured some European wood bison fat into a hollowed flat stone and used an ember from a campfire to ignite it.

He turned and entered a narrow passage in the cliff face behind him, and began to walk deep into a cave system. Led by the flickering flame, his shadows rose and fell against the walls of rock, as he walked, crawled, and climbed for as much as half an hour to reach a vast cavern deep within. There he paused and stared around him at walls adorned with paintings of extraordinary beauty. Some of these were his own work.

Some of them were created by his ancestors, his family, or his friends. For him, the images of non-human animals, humans, and geometric patterns were of immense significance. Slowly, from within his pelt clothing, he drew out a rolled-up package made of birch bark containing a dark pigment. He took some into his mouth, raised his right hand against the wall of the cave, and sprayed the paint over his spread-out fingers.

Gently lowering was a hand, he saw its shape perfectly imprinted on the rock before him. He was satisfied.

Such an action had been carried out many times during the long years that anatomically modern humans had occupied this part of the world, that there were reasons why anatomically modern humans like these painted the insides of caves. Some beliefs urged them to go to such lengths to create art, that the nature of the ceremonies that were central to their work. All across religion at that time, thousands of images were having been painted, depicting the natural world in extraordinary, imaginative, and beautiful detail. The key is that these humans chose to take time from the rigors of physical survival to practice a newly-developed skill that they believed was important to them. It is to marvel at this great explosion of creativity that occurred among these anatomically modern humans and recognized that their minds had reached a level of advancement that was indistinguishable from their later descended generations.

In addition to that, the art and adornments of modern humans also provide a remarkable insight into their way of life. Their stone tools, the beads and shells that they used to create necklaces or pendants, and the fine bone needles they used to stitch their clothing, show the intricate map of the travels that they had undertaken. Shells from the coast of Western Europe were found among the modern human settlements in the mountains to the east, beads from central Europe were found among settlements on the Atlantic coast, seeds of lowland plants were found amongst the groups that lived in upland caves.

Clearly, these humans had built a web of trade and exchange throughout Europe, meeting other groups (or tribes), exchanging goods, and at the same time widening their gene pool, an evolutionary strategy for success.

When the first modern humans reached Europe, they began to coexist alongside the Neanderthals, who by contrast left no signs of trade or travel. All of their materials were local. They had dominated the land for over 200,000 years and their large brains had great ingenuity, to adapt to the succession of climatic and environmental changes that occurred throughout the epoch.

Yet in all that time, the Neanderthals had

created no art, nor had they an imagination as high as modern humans. Some of them buried their dead, but there was no sign of any ritual associated with that, and it could be simply a means of cleanly disposing of a corpse, rather than an expression of belief in an afterlife.

But then, perhaps 500 generations after the time that modern humans entered their world, the Neanderthals too began to express traces of works of art. Simple tooth pendants were found among the layers, in western European caves, left by Neanderthal occupation. These were clearly objects of adornment, with no practical purpose other than display. Perhaps these imaginative leaped to greater sophistication and perhaps survival was simply copied from the practices that they observed and realized were valuable to their ever more powerful neighbors.

Whichever it was, it shall not be enough to prevent the modern human domination of Europe by that time. The modern human ability to network across large distances, along with their creative ability, meant that they could adapt to almost any climate or environmental change, and there had been many fluctuations in the climate throughout the period up to the end of the Last Glacial Period (last ice age). There had been no great violent takeover of the Neanderthals, but rather that the modern humans had simply begun to outcompete them for resources in times of both plenty and harshness.

Living in small groups, the Neanderthal population had become a collection of isolated pockets. And the Neanderthals thought of their new neighbors in some ways quite like them but in some ways very different. But with their far-reaching networks and strange customs, and with so many more of them arriving with each new generation, they could seem quite terrifying.

There is a saying in evolution that you do not have to fail to become extinct; you just have to succeed a little less often. In the end, the Neanderthals would die out sometime around 28,000 BC. And when that moment came, modern humans would become one of the last human species to exist on the planet (the other remaining human species being the 3 to 4 foot tall Homo floresiensis on the Indonesian island of Flores, descended from Homo erectus).

"The Garden of Eden is no more. We have changed the world so much that scientists say we are in a new geological age: the Anthropocene, the age of humans."

—Sir David Attenborough

"We're living in the anthropocene age and now human beings will be the shapers of our future, that totally control the overall functions of not just our planet, but our relationship with other planets."

—Vandana Shiva

PRECURSORS

www.ingramcontent.com/pod-product-compliance
Lightning Source LLC
Chambersburg PA
CBHW060827220526
45466CB00003B/1005